OUR DYNAMIC WORLD: THE

Ordinary Level Students use **TWO** textbooks:	Higher Level Students use
✹ *Our Dynamic World 1* (plus optional workbook)	✹ *Our Dynamic World 1* (pl
✹ *Our Dynamic World 2* **or** *Our Dynamic World 3*	✹ *Our Dynamic World 2* **or** *Our Dynamic World 3*
	✹ *Our Dynamic World 4* **or** *Our Dynamic World 5*

CORE
All students must cover Book 1
(Workbook highly recommended)

Book 1: covers the core sections of the syllabus which must be taken by all students

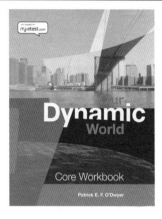

A workbook to accompany *Our Dynamic World 1*

ELECTIVES
All students must cover *either* Book 2 *or* Book 3

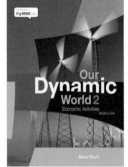

OR

Book 2: Economic Activities – Elective Unit

Book 3: The Human Environment – Elective Unit

OPTIONS
Higher Level only students cover *either* Book 4 *or* Book 5

Higher level only

 OR

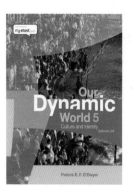

Book 4: Global Interdependence – Optional Unit

Book 5: Culture and Identity – Optional Unit

Our Dynamic World

Global Interdependence (Option)

Charles Hayes

My-etest

Packed full of extra questions, **my-etest** lets you revise –
at your own pace – when you want – where you want.
Test yourself on our FREE website www.my-etest.com
and check out how well you score!

Teachers!

Print an etest and give it for homework or a class test.

GILL & MACMILLAN

Gill & Macmillan Ltd
Hume Avenue
Park West
Dublin 12
with associated companies throughout the world
www.gillmacmillan.ie

© Charles Hayes 2004
0 7171 3587 X
Design, illustration and print origination in Ireland by Design Image, Dublin
Colour reproduction by Ultragraphics Ltd, Dublin

The paper used in this book is made from the wood pulp of managed forests. For every tree felled, at least one tree is planted, thereby renewing natural resources.

All rights reserved.
No part of this publication may be copied, reproduced or transmitted in any form or by any means without written permission of the publishers or else under the terms of any licence permitting limited copying issued by the Irish Copyright Licensing Agency, The Writers' Centre, Parnell Square, Dublin 1.

SYLLABUS OUTLINE

OPTIONAL UNIT: GLOBAL INTERDEPENDENCE

	Content description	National settings	International settings
1	**Statement:** Views of development and underdevelopment are subject to change. [chapters 1–3] Students should • challenge all views of development • critically examine contrasting models and approaches to development including: • determinist and modernisation approaches to development • images and language associated with developing societies • a critical examination of the idea of 'first world and third world' north and south • eurocentric thinking.		Appropriate examples from a world region.

2	**Statement:**
	We live in an interdependent global economy. Actions or decisions taken in one area have an impact on other areas. [chapters 4–6]
	
	Students should study
	• a case study of a specific multinational company with reference to the impact of global trading patterns in relation to both producer and consumer regions
	• the global environmental issues of
	  – deforestation
	  – desertification
	  – global warming
	• the impact of social and political decisions, including
	  – economic and political refugees
	  – migration patterns
	  – human rights issues.
		Irish trade, MNCs in Ireland.
		
		Immigration of refugees.
			Appropriate examples from a continental or sub continental region.
			
			European examples, e.g. managed forests in Scandinavia.
			
			European examples.

OPTIONAL UNIT: GLOBAL INTERDEPENDENCE (continued)

	Content description	National settings	International settings
3	**Statement:** **Empowering people is a way of linking economic growth with human development.** [chapters 7–11] Students should study • the weight of national debt and its impact on the cycle of poverty • the 'aid' debate. Who benefits? • the role of NGOs • land ownership patterns and their impact on development • decision-making processes and levels of participation • levels of exploitation at local and global scales • differing gender roles in society.	Irish aid programmes. Irish NGOs. Nineteenth-century Ireland. Co-operatives. Local enterprise boards.	EU aid programmes. European examples of co-operation. Appropriate examples from a continental or sub continental region.
4	**Statement:** **Sustainable development as a model for future human and economic development.** [chapters 12–15] Students should study • the sustainable use of resources • the goal of fair trade and its potential impact on development • justice issues, particularly in relation to minority groups • the idea of self reliance – development as self-help.	Irish examples.	Appropriate examples from a continental or sub continental region.

Contents

	Preface	ix
	Acknowledgments	x
1	Models of Development	1
2	Images, Language and Eurocentric Thinking	11
3	Describing Divisions – Examining Labels and Models	17
4	Some Impacts of Transnational Corporations	23
5	Deforestation, Global Warming and Desertification	29
6	The Movers	40
7	International Debt and Cycles of Poverty	52
8	The Aid Debate	57
9	Empowerment through Land Ownership, Co-operation and Participation	67
10	Human Exploitation	72
11	Gender Roles	77
12	The Sustainable Use of Resources	85
13	Fair Trade	92
14	Justice and Minority Groups	97
15	Self-Help Development	104
	Picture Credits	109

Preface

This book fully meets the demands of Optional areas of study, which must be undertaken by all students aspiring to higher level Leaving Certificate Geography.

On choosing from the Options available, I have had no hesitation in writing on **global interdependence**. This subject covers a wide range of important and urgent issues that are topical to today's world. A sample of such issues includes global warming, gender issues and international debt on a global scale, together with topics related to immigration and the situation of the Traveller community on a national scale.

This book is designed to help our students to score the best possible grades in Leaving Certificate Geography. I envisage, however, that the study of **global interdependence** will offer much more than a key to high examination grades. I am confident that exposure to the topical and urgent issues contained in this Option of the syllabus will encourage our students along paths of increased personal growth and social awareness.

Charles Hayes, M.A., M.E.d., H.D.E.
St Mary's High School
Midleton
County Cork

Acknowledgments

The author is deeply appreciative of the many people who assisted in the completion of this work. My special thanks to all who took part in the preparation, design and production of the text, especially to Tess Tattersall, Hubert Mahony, Helen Thompson, Regina Barrett and Jamie Biggins (the staff at Gill and Macmillan), Kristin Jensen and Peter Duffy, as well as the staff at Design Image. Thanks also to those of my fellow geography teachers whose valued opinions and advice have had important inputs in this work.

The assistance of the following are greatly appreciated: Baby Milk Action, Bord na Móna, Jacinta Brack, Sandra Carroll, Citizen Traveller, Comhlamh, Corporate Watch, Department of Foreign Affairs, Department of Justice, Department of the Marine and Natural Resources, Una Dunne, *The Ecologist* (London), East Cork Area Development Ltd, Eugene Fraser, International Baby Food Action, *The Irish Examiner*, *The Irish Times*, Irish Traveller Movement, Catherine Joyce, Anne Kinsella, Tom McGrath, *Multinational Monitor* (Boston), *New Internationalist*, Finbarr O'Connell, Kevin O'Dwyer, Crisse O'Sullivan, Pavee Point, Kathleen Regan, Conor Ryan, Jill Ryan, Richard Wilson, *Time Magazine*, Traveller Visibility Group, Trócaire and Wyeth Nutritionals Ireland.

CHAPTER 1
MODELS OF DEVELOPMENT

What Is Development?

Almost everybody is in favour of development. People may disagree strongly, however, on what development actually means.

Some people feel that development is mainly concerned with continued economic growth for the individual. They tend to measure development in terms of the wealth produced and usually measure this in terms of **gross national product (GNP)**.

More and more people now point out that real development must also concern itself with issues other than personal economic growth, stating that real development must help *all* of society, especially those who are poor and weak. It must foster spiritual and cultural as well as economic growth and it must also be **sustainable**. This means that it must not damage the environment or use up the world's resources in such a way that would interfere with the future development of our planet.

It is clear that true *development must serve the real needs of humanity*. Psychologist Abraham Maslow has suggested that there is a '**ladder of human needs**'. The lowest step on this ladder contains the most important human needs, which are basic physical needs. Malsow believed that only when basic human needs are satisfied will people be able to ascend to the second step of the ladder, which deals with the need for security. Then, when people achieve security, they might ascend to the next step of the ladder, and so on. Maslow's ladder of human needs is one useful way of measuring the effectiveness of development (see Figure 1.1).

5. Personal growth
This need can be attained through the development of educational, artistic, sporting, musical or other talents. It is very difficult to meet these 'higher' developmental needs unless the lower levels of the 'ladder of human needs' have all been met.

4. Self respect
This need might be achieved through secure and valued employment, an adequate income, and the realisation that each person is a valued member of society.

3. Love and belonging
People need to feel personally loved by family members or other individuals. They also need to feel valued by members of larger groups, such as peer groups, teams, schools, communities and nations.

2. Security
People need protection against violence by any human being, by the state or by any other group of people.

1. Basic physical needs
All people need secure supplies of clean water, adequate and well-balanced diets, adequate shelter and access to basic health care facilities. The very basis of development is that all people should have access to these basic physical needs. Unless a person can satisfy these needs, he or she is unlikely to achieve the 'higher' levels of human needs.

Fig. 1.1 Maslow's ladder of human needs.
1. Use Maslow's ladder of human needs to consider the following:
   (a) In terms of human needs, how well developed is modern Ireland? Remember that these needs must apply to *all* people living in our land.
   (b) How well off are *you* in terms of human needs?

DEVELOPMENT MODELS

Throughout time, various people and nations supported and practised different approaches to development. These approaches are sometimes called 'development models'. Each model has its own strengths and weaknesses, and each should be questioned on the basis of how well it promoted human development. The following is a sample of some contrasting models of development.

Model 1 – National Self-Reliance

This model was common between the 1940s and 1980s. It was once practised in Latin American countries such as *Peru* and is still practised to some extent in countries such as *India*.

A Bord na Móna briquette factory at Littleton Briquette Factory in Tipperary.
During the decades following Independence, Irish governments encouraged national self-reliance and set up many state-owned and semi-state enterprises, such as Bord na Móna, Irish Shipping and Erin Foods Limited. These industries and services provided Irish-made products and much-needed employment. Several of these enterprises have now closed or been privatised.

Theory

The **general aim** of this approach to development was to balance export earnings against import costs and to avoid the dependency on foreign capital, markets and aid as much as possible.

An important guideline of this model was that the nation should try to meet as much of its production needs as possible. The state would take over (**nationalise**) key industries and would also practise **protectionism**. This means it would place import-tariffs (taxes) on competing foreign goods in order to support home industry.

Good Points

The positive results of this development model soon showed themselves. Third World countries developed a sense of pride and **self-belief** in their ability to improve themselves. Profits from state-run industries were used not for personal profit, but to help create better standards of health, education and other **social services**. More **wealth** was also initially created. The GNP of Mexico, for example, tripled between 1950 and 1970.

Weaknesses

By the mid-1970s, however, serious problems arose for the national self-reliance model:
- National **markets** were usually too small or too poor to allow the continued development of national industries. Foreign markets were difficult to penetrate because of foreign import-tariffs.
- Protected state industries became monopolistic. They were subject to no real competition and so had little incentive to remain efficient and keep costs down. The products and services of some companies gradually began to decline in quality. Other companies allowed themselves to become loss-making in the knowledge that the state would not allow them to become bankrupt.

- The **oil crisis** of the 1970s seriously damaged the economies of countries such as Mexico and India and opened the way for them to accept large loans from international banks. Rising interest rates made the payment of these national **debts** increasingly difficult and forced debtor countries to seek help from organisations such as the World Bank and the International Monetary Fund (IMF). These American-dominated organisations agreed to ease the debt payments of the debtor countries, but only on condition that they 'structurally adjust' their economies and development models. The '**structural adjustment programmes**' opened up economies to foreign competition. This trend towards **economic globalisation** brought an end to national self-reliance as a development model in almost all countries.

> For a more detailed account of the effects of national debts, see pages 52–54.

Model 2 – Centrally Planned Development

This development approach was broadly based on the Communist ideas of Karl Marx. It was originally developed in the former USSR (Russia) and China, but was practised in various forms throughout the 1960s and 1970s in a number of Third World countries. Central planning is still practised (with recent modifications) in *Cuba* and *China*.

Theory

The **aim** of this approach to development is to create **greater equality** in the distribution of national wealth and to provide for the **principal human needs** of every citizen. It follows Marx's belief that the state should give 'to each according to his needs' and receive 'from each according to his ability'.

This model seeks to develop an economy by means of **planned industrialisation**. Factories, land, banks and other commercial entities were **nationalised** and managed by the state on behalf of the people. Profits were invested in **social services** and **producer goods** (such as steel) for further production, rather than in consumer goods and personal 'luxuries'.

Cuba in 2004.
Centrally planned development in this Third World island-nation has not stimulated the production of a great deal of personal wealth. It has, however, succeeded in distributing wealth to provide high standards of education and other social services for all.

Good Points

The positive results of many centrally planned states lay mainly in **social gains** for all people. Cuba, for example, enjoyed tremendous improvements in health care, education, the equality of women and sporting achievements following the socialist revolution led by Fidel Castro. In the socialist-controlled Kerala region of India, over ninety per cent of girls as well as boys receive a formal education and the average life expectancy is nineteen years higher than in the rest of the country.

Weaknesses

Problems associated with central planning include the **discouragement of individual initiative and political choice**. Strict government control over production severely limited private enterprise. Some monopolistic state industries began to produce shoddy goods or provide inefficient services. Consumer and 'luxury' goods, not being widely produced, were generally in short supply. Each of these factors combined to severely limit economic growth. In politics, too, individualism was frowned upon, leading to rather rigid one-party political systems. Largely because of its **failure to bring about prolonged economic growth**, this model has been abandoned in most of the countries that once practised it.

Model 3 – Modernisation and Free Enterprise

At present, this development model **dominates** most of the world. Typified by the *United States*, it now holds sway in a great number of countries ranging from *Brazil* to *Ireland* to *Nigeria*.

The Irish economy has grown enormously as part of a modernised global economy. Many modern factories, such as the computer plant shown here, are owned by foreign multinational companies. They produce high-value, low-bulk products for export.

Theory

This model's aim is to provide a **global economy** in which all barriers to trade will be removed and in which **private enterprise or capitalism** would be allowed to operate with minimum interference from governments or people. The theory of this model is that countries should make **exports** rather than self-reliance the priority for development. Wealth created by exports could be used to pay for imports and (especially in the case of Third World countries) to pay back international debts. The theory of this model is that free international competition will result in continued economic growth. The benefits of economic growth will eventually 'trickle down' even to the poorest people, making everybody better off than before.

Good Points

The principal positive result of this approach to development has been the **massive creation of wealth**. The world's leading economy – the USA – is the cornerstone of capitalist development. The Irish economy, too, has grown enormously since the days of national self-reliance in the 1940s to the present period of an export-driven economy. The economies of some Third World countries, such as South Korea and Brazil, have also grown significantly. A minority of people have amassed great personal wealth. A vast variety of consumer and luxury goods and services are readily available for those who can purchase them.

Criticisms

The modernist, free-market approach has, in its purest form, given rise to several concerns, especially of a social, cultural or moral nature:

(a) This model tends to value **profit above people**. In developing countries especially, some free-market wealth has been created only through the abuse of women and children who through poverty are forced to work in very bad conditions for low pay (see photograph).

(b) The free-market model emphasises the creation of wealth but pays little heed to its fair **distribution**. This has contributed to grave and widening economic inequalities, both between and within countries. The world's seven richest men now own more personal wealth than it would take to provide basic water supplies, health and education facilities for all the poor of the world. Meanwhile, the poorest twenty per cent of the world's population receive little more than one per cent of total world income (see Figure 1.2).

A young child worker in an Asian carpet factory.
The exploitation of child labour has been a feature of free enterprise development in some parts of the world.

(c) The modernist model tends to discourage state involvement in running social services such as public transport, water supplies and even public health. It favours instead the privatisation of public services so that they can be run 'efficiently' by private companies for profit. In countries such as the United Kingdom, privatisation has resulted in the decline of many public services. This has adversely affected the quality of life of many poorer people and so has led to increased inequality between rich and poor. In developing countries, the effects of '**public services cutbacks**' have been even more dramatic. Latin American countries such as Brazil now make state-of-the-art sports cars while schools in working-class areas are crumbling.

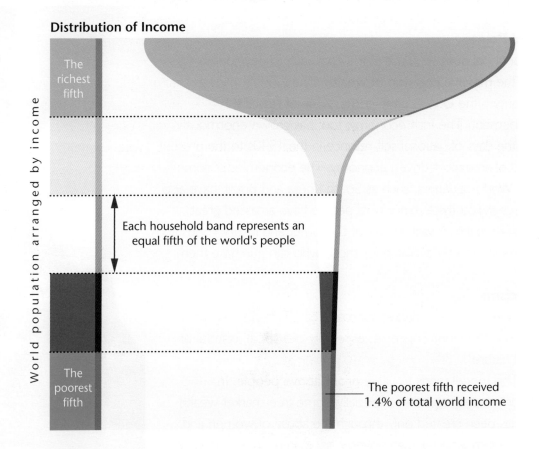

Fig. 1.2 Describe what this diagram reveals about the distribution of income in the world.

(d) This model of economic modernisation was semi-imposed on countries such as Russia, Nigeria and Mexico as part of the **structural adjustment programmes** prescribed by the International Monetary Fund. These programmes, while often increasing the GNP of the countries involved, brought many hardships for workers and poorer people:

- *Unemployment increased* dramatically as formerly state-supported industries were 'slimmed down' to make more profits for their private owners.
- Enforced devaluation of local currencies made exports easier to achieve, but they also greatly increased the costs of imported oil and other essentials. This caused *increases in the costs of basic goods,* which proved disastrous for the poor and unemployed.

- The governments of the developing countries were forbidden to subsidise the price of basic foodstuffs because this might hinder 'international free trade'. The resulting *increases in the costs of basic foods* pushed many to the brink of hunger. In Nigeria, the percentage of severely malnourished children increased by thirteen per cent in six years.
- Third World countries were encouraged to greatly increase exports so that they could better pay their debts to foreign bankers. Nigeria, for example, was encouraged to give more and more land over to the production of cocoa. This resulted in less basic food being grown for local consumption and so led to increased malnutrition. Furthermore, so much cocoa was produced that the international market became saturated, causing the price of cocoa to fall sharply. Nigeria ended up getting less money for more exports!

(e) The modernisation approach can be described as being **culturally imperialistic** in nature. It assumes that traditional rural societies are economically 'backward' and that countries such as the United States are the models of 'development'. The approach assumes that in order to attain development, Third World countries must automatically follow the social and economic paths taken by the United States or Western Europe.

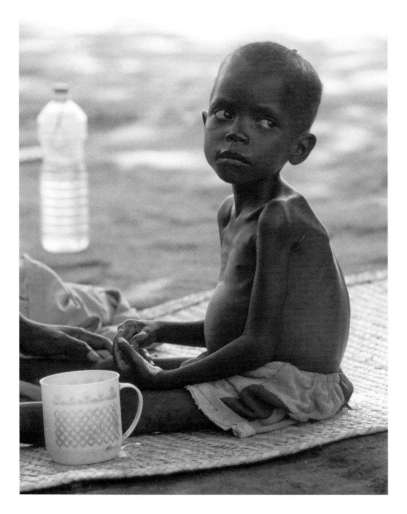

Malnourished children in Nigeria have increased by thirteen per cent in six years. Explain how **structural adjustment programmes** have contributed to this increase.

(a) What points do the **cartoons** on this page make about development?
(b) Do you think that the messages given by the cartoons are fair and balanced? Explain your views.

With regard to the news extract below:
(a) Define in your own words the concept of 'appropriate development'.
(b) Give some examples of 'pseudo-development' and 'cultural imperialism' other than the examples given in this extract.
(c) Do you agree that 'appropriate development' is desirable and achievable? Explain your point of view.

On the Need for Appropriate Development

Perhaps what really matters in development is for the people of an area to be able to meet their own needs, solve their own problems, guarantee the ecological survival of the area and enjoy life in a satisfactory state and at a satisfactory pace. This is what we might call appropriate development, because it is appropriate to the real needs of the ordinary people in any given area.

Neither capitalist, socialist or national self-sufficiency models appear to be fully appropriate to real development. The capitalist model puts profit first. It promotes pseudo-development with the over-consumption of unnecessary goods, such as the aspiration for every teenager to own a mobile phone. It promotes the growth of cultural imperialism, such as the consumption of Coca-Cola in rural Africa or the wearing of jeans by squatters living on the garbage tip in Smoky Mountain in the Philippines. Both the state-socialist and the national self-sufficiency models emphasise grand national plans to the detriment of the individual and of the all-important local communities.

Appropriate development requires real co-operative community action as well as local control and participation. With appropriate development, the individual is liberated from the tyranny of endless competition and from the bureaucracy of the state. He or she is free to develop in the only way that is meaningful for human beings — as part of a tightly knit yet economically and socially viable group.

Views on Underdevelopment

People's views on underdevelopment have changed over time. Until recently, most Westerners assumed that underdevelopment was **a natural state** in communities that had failed to 'develop' along Western European or North American lines. This *European or Eurocentric* view portrayed the typical characteristics of underdevelopment as being low GNP, a prevalence of subsistence farming, high birth rates and a general absence of Western values and customs. Even school textbooks reflected the idea that the spread of Western influences was the main avenue through which 'primitive' societies could eventually attain development.

Many people now believe that underdevelopment is a **situation arrived at** rather than a natural state. They point out that underdevelopment in some societies is directly connected to the economic development of other societies – that the wealth and power of some depends on the exploitation and poverty of others. There is much evidence to support this view that development and underdevelopment can be the opposite sides of the same coin.

In the seventeenth and early eighteenth centuries, India was one of the world's most economically developed countries and was famous for its exports of fine textiles. On being taken over by Britain, India's manufacturing industry was discouraged in favour of British manufacturing. India became a producer of cheap raw materials and a Third World country.

When Spanish adventurers invaded Central and South America they found Mayan and Inca civilisations that were highly advanced. They encountered highways, bridges, cities and agricultural practices more advanced than those in Europe at that time. Then the Incas and Mayans were enslaved by the Spanish and their civilisations were destroyed. Their descendants in Guatemala and Peru now form part of the Third World poor.

Many sub-Saharan African peoples, such as those in present-day Tanzania and Zimbabwe, suffered the loss of their lands as well as their independence to European settlers during the nineteenth and twentieth centuries. As such regions became European colonies, much of their best lands were used to grow luxury crops for export in a global economy rather than food for local consumption. As local people lost their self-sufficiency in food production, hunger, famines and migrations to urban shanty towns became common occurrences. Such colonialism brought power and wealth to some, but subservience and poverty to many.

At the present time, large multinational companies transfer vast amounts of wealth to their home-base countries in America, Western Europe and Japan. But some of this wealth is generated by the labour of poor Third World people, who receive very little for their contributions to global economic development.

Activities

1. (a) Outline the advantages and disadvantages of modernisation and free enterprise as a model of development.
   (b) Describe a model of development other than that of modernisation and free enterprise. Describe the advantages and weaknesses of the model you describe.
   (c) Write a paragraph describing each of the following terms:
   - Maslow's ladder of human needs
   - Structural adjustment programmes
   - Appropriate development.
   (d) 'Underdevelopment is ... a product of history. It is not an original condition of man, nor is it merely a way of describing the economic status of a "traditional" society. Underdevelopment is part of the same process which produced development'.
   – Historian Keith Griffin
   To what extent do you agree with the above statement?

2. (a) What is meant by each of the following terms: *deforestation*, *salination* and *desertification*?
   (b) Discuss the overall message of the cartoon shown below.

Don't you see, he'll go on sprouting more and more heads. It's the body itself you must go for.   The Ecologist.

CHAPTER 2
IMAGES, LANGUAGE AND EUROCENTRIC THINKING

EUROCENTRIC THINKING

With the rise of the British, French and other European empires, Europeans were presented with the notion that they were more 'developed' and therefore somehow superior to the peoples of European colonies. Some European newspapers, administrators and colonists began to refer to colonised peoples insultingly as 'natives' and inaccurately as 'savages'. Local cultures and languages were frequently portrayed as being inferior and insignificant.

Many Europeans, influenced by these messages, began to think **Eurocentrically**. They began to regard Europe as the cultural and developmental centre of the world. Some adopted hostile and blatantly racist attitudes towards colonised peoples. Others came to believe that it was 'the white man's burden' to 'civilise' colonised peoples by encouraging them to adopt the languages and cultures of their European conquerors. Many simply believed that the European way of life was the 'natural' one, and that it ought to be copied in other parts of the world. Eurocentric thinking became a lasting obstacle to the development of real understanding and mutual respect between the peoples of the First and Third World.

Messages of supposed European superiority were also presented in more subtle forms. Up to recent times, the principal wall map used in European schools has been **the Mercator Projection**, from which European children learned the locations of places throughout the world. But the Mercator map (while accurately showing the *shapes* of land masses) presents a very **Eurocentric** view of the world. It exaggerates the importance of Europe by placing it at the centre of the map and so, by inference, at the centre of the world. The Mercator projection also exaggerates the *sizes* of places in high and mid-latitudes in comparison to places in low latitudes. European children thus began to believe that their own mid-latitude countries were much bigger on a world scale than is really the case (see Figure 2.1).

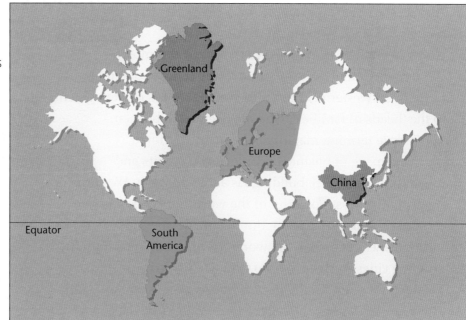

Fig. 2.1 The world according to Mercator. The following points illustrate some of the misleading Eurocentric messages portrayed by the Mercator map:
- **Fact:** China is about four times larger than Greenland. *(Which seems to be larger on the map?)*
- **Fact:** South America is twice the size of all of Europe. *(Which seems to be larger on the map?)*
- **Fact:** Half of the world lies south of the Equator. *(What does the map suggest?)*

Are *You* a Eurocentric Thinker?

When you think of the word 'farmer' do you think of a man? When you think of 'business people', do you think in terms of people in smart suits with briefcases in office buildings? If the answer is yes to either of those questions, you may be suffering from a bout of Eurocentric thinking.

The fact is that the majority of farmers in the world are women. In the Third World, where most of the world's farmers live, women carry out most agricultural work. In most traditional African societies, for example, men clear and plough the land. Women, on the other hand, plant, weed, hoe, reap and save the crops. But most Westerners cannot quite get used to the concept of female farmers, so Western aid to Third World farmers is often given to the 'heads of families', who are deemed to be men. The female farming majority is frequently ignored.

The image of business people in smart suits and inhabiting large office buildings is another largely Western concept. The most powerful of the world's business sector certainly fit this description. But the image does not extend to most business people in the world's largest and most rapidly growing cities, such as Mexico City, Sao Paolo or Calcutta. There you will find millions of business people without suits. These are the tailors, the sellers, the repairers, the cleaners, the shoe-shiners and a host of other self-employed entrepreneurs who make up the 'informal' business sector of business. The wealth that these people generate is not counted by us Europeans as part of a country's GNP, yet these 'barefoot business people' are as much a part of capitalism as is any European in a pin-striped suit.

A female welder in a car workshop in Bangladesh.

A typical African farmer.
How do the images above differ from Eurocentric images of workers?

We now no longer refer to Third World peoples as 'natives' or use the Mercator projection exclusively in our schools, yet the **images that we receive of Majority World peoples are often gravely misleading**. This can be a serious matter because our relationships with people in the Majority World depend largely on the images and language with which they are portrayed to us. These images will largely determine whether we see people in the Third World as dignified members of our human family in search of justice or as pitiable (and perhaps slightly inferior) creatures worthy merely of charity.

MEDIA IMAGES OF THE THIRD WORLD

Most Irish people rely on television, newspapers and radio reports for information on world affairs, yet the following facts show that our media can sometimes present us with inadequate and distorted views of people and events in Third World countries.

- The Majority World contains more than two-thirds of our human family. It receives **less than one-tenth of all Irish news coverage.**
- Wars, famines and other disasters directly affect the daily lives of a very small proportion of Third World inhabitants, yet almost all our Third World **news coverage focuses on** such **disasters**. Bad news sells news items.
- Where disasters do happen it is usually local people who do most to alleviate the hardships caused, yet local relief efforts are often under-reported and most of our news items emphasise the importance of Irish or other foreign emergency aid in alleviating suffering. **Local people are often inaccurately stereotyped as inactive, helpless victims** who are dependent on Westerners for survival. This image can reinforce prejudice and attitudes of superiority among us towards peoples of the Third World.
- **News items often trivialise** even the most serious of disasters, seeking 'human interest stories' in the midst of appalling human suffering. In the late 1990s, serious flooding devastated large areas of Mozambique. These floods killed thousands of people and made hundreds of thousands homeless, yet many TV and newspaper images focused attention on a young Mozambican woman who, fleeing the floods, gave birth to her baby in a tree before being rescued by a foreign helicopter crew.
- Most news items are **shallow and unquestioning**. Few reports investigate the root causes of Third World poverty – causes that often relate to unjust world trading systems from which First World countries benefit.
- Even 'serious' news items often depend on visiting journalists and Western 'experts' for commentaries on complex local Third World issues. They **seldom seek the views of local people**, who are usually the real experts on their own affairs.

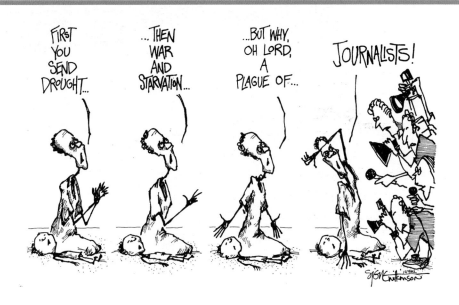

Aid Agency Images

Irish and other nongovernmental agencies (NGOs) do a great deal of excellent development work in Third World countries. Many such agencies also work hard to present us with accurate images of the Majority World.

However, NGOs may sometimes feature images of starving children or other suffering people in their appeals for public subscriptions. The advantage of these images is that they stimulate human pity and so entice people to contribute to the alleviation of poverty. On the other hand, such images can present the following problems:

- They may reinforce a stereotypical view of Third World people as helpless and dependent on Westerners.
- They may portray the message that 'charity' rather than justice is the answer to developmental problems.
- If repeated too often, such images tend to have diminishing visual impact. This may lead to 'donor fatigue' and reduced donations.

Study the photographs, each of which could be used to advertise the work off aid agencies.
(a) Contrast the images of Third World people presented by these photographs.
(b) In the case of each picture, write a paragraph to describe the Third World people shown. Choose your language carefully to suit the images given in each picture. Use adjectives such as some of the following: *lazy, greedy, dependent, hard-working, oppressed, pathetic, successful, independent, helpless, powerful, weak, noble, clever, inferior, equal, superior, capable, bright, dull*.
(c) Should aid agencies ever produce advertisements featuring starving people? Explain your point of view.

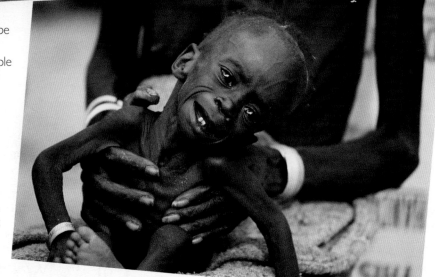

Language Associated with Developing Societies

Most people are **positive, rational** and **informed** in their approaches to development and developing societies. Others can be **irrational, condescending, uninformed, Eurocentric, racist** or **openly hostile**.

Attitudes towards development and towards developing societies are often revealed in the type of language used in referring to such societies. The following is a selection of quotations relating to development. *From the quotations identify some examples of each characteristic listed in bold type in the previous paragraph.*

1 '(British Rule) has impoverished the people (of India) by a system of progressive exploitation and ruinously expensive military and civil administration which the country can never afford. It has reduced us politically to serfdom. It has sapped the foundations of our culture.'
- M. Gandhi in 1930.

2 'While the countries of Western Europe have to a large extent outgrown nationalism ... many countries in the rest of the world have only just become conscious of nationalism; the emergent countries of Africa, the Middle East and Latin America are all going through the early stages of nationalism.'
- H. Robinson, author of Human Geography, 1969.

3 'I have a duty to help save the heathen from damnation, to wean them from their perversity and to open the gates of a Christian civilisation.'
- A Scottish mill girl who became a missionary in 1870.

4 'The partition of Africa was, as we all recognise, due primarily to the economic necessity of increasing the supplies of raw materials and food to meet the needs of the industrial nations of Europe.'
- Lord Lugard of Britain in 1853.

5 'When we say we're against deforestation, people say we're against the development of Brazil. We're not against development, but we are against the devastation of Amazonia. We want development that doesn't only benefit big companies and the powerful, but the people who work the land.'
- Chico Mendes, leader of Brazilian Rubber Tappers in 1985.

6 'C____ plans to open a nursery for white kids only. She is very unhappy that there are two black children at her daughter's nursery school. She does, however, enjoy the services of a black maid. "I don't let her cook though," C____ explains, "I can't stand the thought of her scratching herself and then touching the food."'
- A right-wing South African interviewed in The Guardian Weekly, January 1994.

7 'The best and most intelligible interpretation (of the term "underdeveloped") may be taken as the condition in which per capita real incomes in a country are low ... Defined in this sense, underdevelopment is synonymous with "poor" ...'
- H. Robinson, author of Human Geography, 1969.

8 'It would not have been possible for Italy to watch, idle and indifferent, the peaceful crusade undertaken by all the other great powers in order to civilise the populations of Africa, without seeing Italy's good name disgraced in Europe.'
- The Italian Foreign Minister referring to Italian colonial expansion in Africa in 1884.

9 'It is my belief that you are as close to God as I am.'
- Irish person on being introduced to an African immigrant, 1998.

10 'In 1492, Christopher Columbus discovered America'.
- A junior school textbook.

Activities

1. (a) What is Eurocentric thinking?
   (b) How might the slave trade and European colonialism have contributed to the rise of Eurocentric thinking? How might Eurocentric thinking have contributed to colonialism?
   (c) What effects might Eurocentric thinking have on human development and on harmony between peoples?
2. Examine the newspaper images in the photograph and cartoons.
   (a) What attitudes are portrayed in each of the images?
   (b) Which of those images do you think might be most offensive to people in Africa and which might be least offensive? Explain.
   (c) *'Images such as those are harmful to people everywhere.'* Discuss.
   (d) In your opinion, why do European newspapers publish images such as those shown?
   (e) Identify ways other than those illustrated by the pictures shown here in which the European media can present distorted images of people and events in Third World countries.

"HOW SWEET... THEY STILL GET THEIR FOOD FROM FIELDS."

CHAPTER 3
DESCRIBING DIVISIONS – EXAMINING LABELS AND MODELS

We are aware that great divisions exist between the rich and poor countries of the world. Vigorous debates exist on how to most conveniently and accurately describe these broad divisions.

The Three-World Model

Up to the 1980s, people described inequality between countries in terms of a three-fold division of the planet into the First World, Second World and Third World (see Figure 3.1):

- **The First World** was made up of countries in Western Europe and North America and also included Australia, New Zealand and Japan. These were the rich countries with capitalist, free-market economies.
- **The Second World** was comprised of communist and socialist countries with centrally planned economies. These consisted mainly of the Eastern European states, such as Poland, Hungary and the former Soviet Union. While not nearly as rich as many First World countries, they generally enjoyed adequate diet, good life expectancy rates and good social services.
- **The Third World** included most countries in Africa, Latin America and Asia. In general, these were the poor countries of the world. The Third World contained seventy-five per cent of the world's people, but only twenty per cent of the world's wealth.

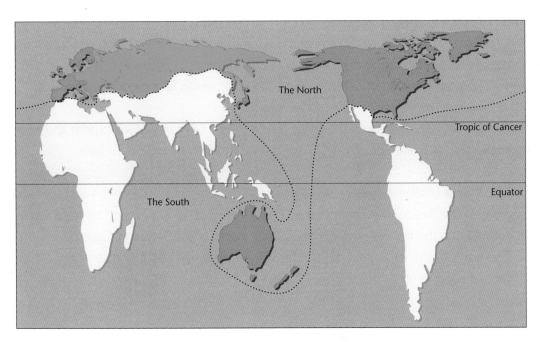

Fig. 3.1 shows the First, Second and Third Worlds. It also shows the world divided into 'North' and 'South'.
(a) This map is known as the **Peter's Projection**. It shows the correct sizes but not the correct shapes of landmasses. Contrast this map with the Mercator map shown on page 11.
(b) This map is similar to one produced in a textbook for schools in Australia. In what way does its 'world view' contrast with that presented by most maps used in Ireland?
(c) Make three separate lists of ten countries in each of the First, Second and Third Worlds.

Several **objections** have been raised to the continued use of the division of our planet into the 'three-world' model:

(a) Many people see it as a product of **Eurocentric** and Western thinking. They see the three-fold division as unacceptable on the grounds that it implies some kind of pecking order of priority or superiority between the world's nations, with our Western societies being thought of as having achieved the highest rates of human development.

(b) The Second World, as described in the three-world model, no longer exists since the collapse of communism in Eastern Europe in the early 1990s. This makes the model **obsolete in general terms**.

(c) Many people object to the term '**Third World**' because they feel it can be equated in people's minds with 'third rate'. They also point out that the term lumps together a wide range of very economically diverse countries. Ethiopia and Brazil, for example, are each considered to be within the Third World, yet Brazil far exceeds Ethiopia in terms of GNP and other developmental criteria. Oil-producing Middle Eastern countries such as Saudi Arabia and Kuwait are now among the world's richest countries, yet they have traditionally been included within the Third World, since the three-world model was conceived before the great oil price rises of the 1970s.

It should be pointed out, however, that the term 'Third World' was never intended as an economic category. It was conceived as a political category and is favoured by many Third World writers and activists. The term draws its inspiration from the time of the French Revolution. In those days, a political grouping called the 'Third Estate' contained the vast majority of French people, but enjoyed little political power or influence. The term 'Third World' **refers to the lack of real political power** enjoyed by poor countries in today's world. From that point of view, it is a term that is acceptable to most geographers and political commentators.

The Two-World Model

In the 1970s, an Independent Commission on International Development met to discuss poverty, famine, unfair trading and other global problems. In 1980, this commission issued the **Brandt Report**, so called because the chairperson of the commission was Willy Brandt, a former Chancellor of Germany.

The Brandt Report ignored the three-world model of world development and instead divided the planet into the two-world model of North and South (see Figure 3.1):

- **The North** comprises the world's richer and more powerful countries and contains the First and Second Worlds together. It is referred to as 'the North' because it lies generally on the northern part of the planet, north of the Tropic of Cancer.
- **The South** is the equivalent of the Third World. It is so called because it lies generally to the south of the richer North.

Examine Figure 3.1. What difficulty does it suggest with the terms 'North' and 'South'?

The North-South model has the **advantage** of omitting the now largely nonexistent Second World category. It also emphasises the fact that political and military power were traditionally shared between the First and Second World nations. The model does, however, have the following **weaknesses**:

(1) **Not all northern countries lie in the northern hemisphere**. Figure 3.1 shows that Australia and New Zealand, each part of the North, lie deep within the southern hemisphere.

(2) **The term 'South'**, like the term 'Third World', lumps together a large number of countries that contrast greatly with each other in terms of development. (See objection (c) to the three-world model.)

(3) **The term 'North'** also embraces a diverse collection of countries that contrast greatly in terms of development. It combines powerful countries such as the United States with new, politically struggling countries such as Bosnia. It equates very rich countries such as Switzerland with countries such as Russia, the social services and economy of which have grown clearly inferior to some 'Southern' countries such as Saudi Arabia.

Farming in Ethiopia.

A prosperous part of Brazil's capital city, Brasilia.

What do these photographs suggest about terms such as 'First World' and 'Third World', or 'North' and 'South'?

A back street in Naples, southern Italy.

The Problem with All Models

A geographical model is a means of presenting a complicated truth or illustrating a complex situation in a simplified manner.

Models are useful in presenting introductory, easy-to-understand views of situations which might otherwise be difficult to grasp. When presented in the form of a flow chart or other diagram, a model can also provide an immediate visual impact of the basics of a complex situation.

The problem with geographical models is that they tend to impart **generalised information** that seldom fully reflects the truth of any particular situation.

The three-world and North-South views of the world are examples of geographical models. They present generalised pictures of large areas of the world without recognising that every country (and sometimes every region within a country) is unique in its own level and type of development. Ireland and Poland are each part of the North, yet their levels of development are quite different. Argentina is considered part of the Third World. Culturally and economically, however, it is closer to some European countries such as Spain than it is to some Third World countries such as India.

Other Terms Used Over Time by People in Rich Countries to Describe Poorer Countries		
Period	Terms used	Comments on terms used
Pre-1950s	'Primitive', 'Backward'	• Eurocentric, racist in connotation.
1950s	'Underdeveloped'	• Generally accurate from a purely economic viewpoint. • Might be hurtful to peoples of the Third World.
1960s–1970s	'Developing'	• Assumed (often incorrectly) that Third World economies would grow. • Potentially less hurtful to Third World sensibilities. • Patronisingly assumed that developing countries were 'developing' along Western lines.
1980s–2000s	'Less developed'	• More accurate than the term 'developing'. Recognises that poverty exists in all parts of the world and so is a relative rather than an absolute term.
	'Majority world'	• Similar to 'Third World' tag. Recognises that most of our human family lives in poor countries.

Table 3.1 *'First World images and language associated with developing societies have themselves developed over time.'* Discuss with reference to Table 3.1.

Activities

1. (a) Briefly describe the two-world model and the three-world model of development. Which model do you think is more appropriate? Why?

   (b) 'Classifications such as "First World" and "Third World" can be both useful and misleading.' Discuss.

   Or

   'Terms such as "First World" and "Third World" should be examined critically rather than accepted without question.' Discuss.

   (c) Describe the extent to which the information given on the map (Figure 3.2) supports the concepts of 'First World' and 'Third World'. In your answer refer liberally to information provided by the map.

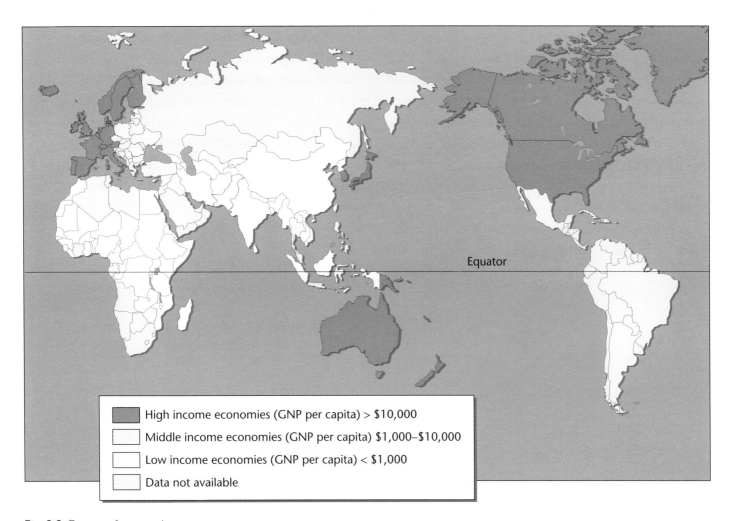

Fig. 3.2 Groups of economies.

(d) One of the problems with concepts such as 'North' and 'South' is that they can reinforce stereotypes and give a false impression of unchanging developmental divides between different peoples.

To what extent is this borne out by the newspaper extract below?

The Choctaw Indians and Ireland

In 1990, Choctaw Indian Chief Hollis Roberts and other Indians from the US came to Ireland and took part in a famine memorial walk in memory of Ireland's Great Famine of 1845–49 as a method of raising money for Somalian famine relief.

The Choctaw Indians were a peaceful race. A Spaniard who met them in 1540 wrote of their hospitality and their nature, which ran completely counter to the 'Red Indian' stereotype beloved of Hollywood. Their civilisation grew near a prehistoric mound in Mississippi.

As European settlers spread across the US, the Choctaw made treaties with France, Britain and the United States.

In 1830, the Choctaw Indians signed the Treaty of Dancing Rabbit Creek. Afterwards, the government, led by President Andrew Jackson, forced them from their land.

The Choctaws began a long walk that became known as the Trail of Tears. They were forcibly removed and made to walk over 500 miles from Mississippi to Oklahoma in the middle of winter. Over half of them died.

In an extraordinary twist, this small Indian tribe developed a link with Ireland in 1847 when news of the Great Famine reached the small Choctaw town of Skullyville. The Indians, who had an intimate knowledge of dispossession and hunger, raised $710, which was sent to Ireland for use as famine relief.

(*The Cork Examiner*, September 1992)

CHAPTER 4
SOME IMPACTS OF TRANSNATIONAL CORPORATIONS

We now live in a global economy in which relatively few powerful companies control more and more of the world's trade and industry. These companies are called **multinational or transnational corporations** (TNCs) because they each operate in more than one country. Over seventy per cent of world trade is now controlled by approximately 500 of these giant companies. Over ninety per cent of TNCs have their headquarters in industrialised countries such as the United States.

Transnationals provide much-needed **investment and employment** in the economies of Ireland and other countries. That is why our government and the Irish Trade Board offer financial incentives to TNCs that set up branch factories in Ireland. Our 'Celtic Tiger' industrial boom owed much of its production to the involvement of overseas-based companies. In fact, branch factories of foreign-based TNCs produce most of our non-agricultural exports.

> Find out more about branch plants of MNCs in *Our Dynamic World 2* Chapter 9.

The world's multinational companies thrive under a system of *global free trade*, which is encouraged and policed by a powerful international body called the *World Trade Organisation*. There is no doubt that free trade has stimulated the creation of a vast array of relatively **low-cost goods**, which services the consumer societies of Ireland and other First World countries.

However, there are serious downsides to the TNC-driven global economy in which we live:

- Multinational companies destroy as well as create employment. Their operations are often capital intensive rather than labour intensive and their arrival can often cause **serious job losses** by putting local firms or small farmers out of business. As some TNCs chase lower and lower production costs in different parts of the world, their creation of employment in one country frequently comes at the cost of job losses in another (see 'The Global Hunt for Cheap Labour' on the next page). The employment that transnationals provide may therefore be quite temporary, as any branch factory can be closed down suddenly by a financial decision made at a far-away TNC headquarters. For example, employees at the Packard multinational plant in Tallaght, County Dublin arrived home from work one evening in April 1996 to learn on the RTÉ news that their factory was to close with a loss of 800 jobs.

(a) Discuss the impacts of the activities described in the text box on the workers and regions involved in the **production** of canned tuna.

(b) Do you think that **consumers** of canned tuna should support the activities described? Explain your point of view.

The Global Hunt for Cheap Labour

Environmentalist Dave Phillips recently told a story about the tuna industry which illuminates the globalisation process.

'In the old days,' he said, 'California had the largest tuna-canning industry in the world, but today the wages in California are about $17 an hour. So the industry moved, first to Puerto Rico, where wages are about $7 an hour and then, when they decided that was too much, to American Samoa, where wages are about $3.50 an hour. From there it moved to Ecuador, where workers are paid about $1 an hour and then on to Thailand, where a great deal of the industry is today and wages are about $4 a day! And now, amazingly enough, there is some movement to Indonesia, where wages are as low as a couple of dollars a day.'

Wages in Indonesia are so low, of course, because union organisation there is weak under military dictatorship and working conditions are therefore poor.

(*New Internationalist*, January/February 1997)

- In some instances, economic globalisation can **undermine** local **democracy**, workers' rights to form trade unions and even national independence. Multinationals are often wealthier than the countries in which they operate and this can make it difficult for national governments to control TNC activities. Shell, which operates oil interests in Nigeria, has annual sales which are three times the total annual income of Nigeria's 100 million people. In the 1990s, a subsidiary of the Chiquita banana company controlled almost three-quarters of vital banana production in Panama.
- Much of the profits made by TNCs return to the corporations' home countries. It is estimated that such '**profit flight**' allows US multinationals to take hundreds of millions of dollars more out of Africa each year than they invest or leave there.

Case Study: Wyeth Nutritionals of Askeaton

Ireland is now Europe's largest producer of infant formula (manufactured baby milk). The country's largest infant formula plant is that of **Wyeth Nutritionals Ireland**, situated near the village of Askeaton, County Limerick (see the photograph and map on the next page).

SOME IMPACTS OF TRANSNATIONAL CORPORATIONS

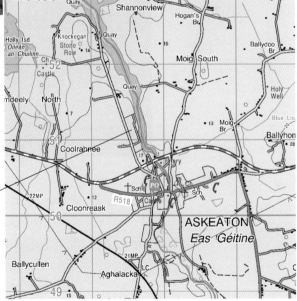

Wyeth Nutritionals of Askeaton.
(a) With reference to the photograph and to the map, describe the *site* and *situation* of the Wyeth plant.
(b) Do you consider the location of the plant to be a suitable one? Explain your answer.

The Wyeth plant at Askeaton is a large branch factory of **American Home Products Corporation**, a transnational corporation with its headquarters in New York (see Figure 4.1).

American Home Products Corporation
(a TNC with headquarters in New York)
▼
Wyeth Nutritionals International
(a subsidiary of American Home Products, with branches and units in many countries)
▼
Wyeth Nutritionals Ireland
(one unit of Wyeth Nutritionals International)

Fig. 4.1 The Wyeth company tree.

25

Impacts on Producer Regions

The establishment of Wyeth Nutritionals Ireland has had significant **positive impacts** on Ireland as a producing region of infant formula:

- More than 600 people are **directly employed** at the Askeaton plant, despite the fact that the plant is capital intensive and automated. The employees can avail of contributory pension schemes, free life assurance and subsidised health insurance. Outside of the factory, indirect employment is fostered in transportation, engineering and other companies that service the Wyeth plant.
- The Wyeth company contributes more than €150 million to the Irish economy each year. Much of this goes to the co-operatives and farmers of Munster's dairy sector. Some goes to Smurfit of Cork, which provides packaging, and to supporting services such as construction, energy and engineering. Some of the wages earned by Wyeth employees are circulated in the Askeaton and Limerick areas. This **multiplier effect** assists the general economies of these areas.

What is meant by the term **'multiplier effect'**? How does the multiplier effect of Wyeth's Askeaton plant influence this creamery in the Golden Vale?

However, it must be remembered that in a world of globalised manufacturing, **the prosperity of one producing area can mean the depression of another**. The main aim of most transnational companies is to seek greater and greater corporate profits for themselves. To achieve this aim, they frequently switch their production units from one geographical area to another. The opening or expansion of a multinational branch plant in one area can therefore be accompanied by the negative impact of closures or decline in other producing areas.

In 1992, Wyeth's Askeaton plant expanded to double its production. Hand in hand with this came the closure of a Wyeth plant in Havant, southern England, with the loss of hundreds of jobs. The expansion of the Askeaton plant in 2000 was similarly accompanied by the closure of alternative factories in Australia and Colombia and of a supply plant in South Africa.

Impacts on Consumer Regions

Economic Impacts

- Wyeth Nutritionals' baby formula products are exported to many parts of the world. These products are **imports** to such countries and, like all imports, must be paid for out of the wealth of the consuming countries. If not balanced adequately by exports, imports can create a **trade deficit** which can ultimately harm the economies of the consuming countries.
- It can also be argued that the imported products of large multinational companies can **hinder the development of alternative products** in the consumer countries themselves. Large multinationals are so immensely wealthy that they can package and down-price their products to stifle any potential competition from within the consuming countries. They may therefore have a negative effect on native economic enterprise within consuming countries.

Fig. 4.2 Describe the causes and effects of the trade situation illustrated here.

Social and Health Impacts

- It is an established fact that breast-feeding is nutritionally and emotionally best for infants. When a mother cannot or chooses not to breast-feed her baby, however, and when sanitary conditions are adequate, baby formula provides a second-best alternative to breast-feeding. It is clear, therefore, that baby infant formula provides some mothers in **economically developed countries** with a convenient though usually inferior alternative to breast-feeding.
- The export of baby formula to **Third World countries** is a controversial issue because the sanitary and economic situations of many families may render the use of baby formula dangerous. The United Nations' Children's Fund (UNICEF) has calculated that every thirty seconds a child dies in the Third World because they are not breast-fed and that Third World babies are twenty-five times more likely to die if they are bottle-fed.

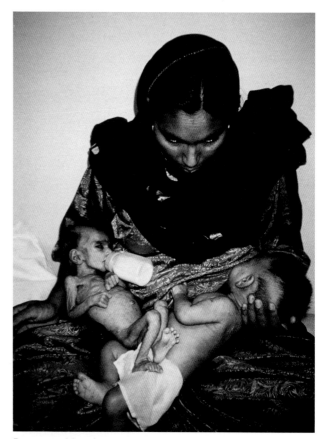

Breast and bottle.
These babies are twins. The baby with the bottle died the day after this photograph was taken, but her breast-fed brother is thriving. This young Pakistani mother was told that she would not have enough milk, so she should bottle-feed the girl. She could almost certainly have fed both twins, because the more a baby sucks, the more milk is produced.
(a) Why do you think the boy was chosen to be breast-fed?
(b) Why do you think the baby girl did not thrive on the bottle?

How Powdered Baby Milk Formula Can Kill in Third World Countries

- The **water** mixed with baby powder can be unsafe and it is often impossible in poor conditions to keep **bottles and teats** sterile. Bottle-feeding in such circumstances leads to **infections** such as diarrhoea, the biggest killer of children worldwide.
- Baby milk is **expensive**, often costing more than half an entire Third World family income. This means that bottle-feeding could contribute to **family malnutrition**. Furthermore, poor mothers trying to make the powder go further may be tempted **to overdilute** the powder. This would result in the malnourishment of the baby.
- Some Third World mothers, thinking that bottle-feeding is a **'modern' alternative** to breast-feeding, may be tempted to feed milk formula to their babies on a trial basis. When a mother does not breast-feed, she will soon stop producing her own milk. She will then have no alternative but to continue trying to feed her baby on expensive artificial milk.

(a) What point is being made by this cartoon?
(b) Do you think the cartoon is fair?

By 1982, the **World Health Assembly** became so concerned about the dangers of baby milk formula in the South that it banned the direct advertising of the milk formula in Third World countries. Wyeth Nutritionals continues to export milk formula to Third World countries, though it now carries a 'breast is best' slogan on its packaging.

Activities

1. Define each of the following terms:
   - Multinational company (MNC)
   - Profit flight
   - Multiplier effect
   - Global economy.
2. In the case of one MNC which you have studied, show how decisions taken in one region can impact positively and negatively (i) on producer regions and (ii) on consumer regions.
3. Attempt to explain how powdered baby milk can endanger health in Third World countries.

CHAPTER 5
DEFORESTATION, GLOBAL WARMING AND DESERTIFICATION

In the previous chapter, which related to the manufacturing industry, you learned that **in an interdependent world, actions or decisions taken in one area often have impacts on other areas**. In this chapter, you will learn that this also applies to the following environmental issues:

- **Deforestation**: the large-scale cutting of forests, such as in the Amazon Basin.
- **Global warming**: the gradual increase in the temperatures of the earth's surface and atmosphere.
- **Desertification**: the gradual spread of desert conditions, such as in the Sahel region in Africa.

A Geographical Model

A geographical model represents a simplified version of a complex geographical truth. A model can be presented in any one of a variety of forms. It might be a diagram, a graph or a statement. Its usefulness lies in its ability to focus on key points while omitting less central details. Its weakness lies in its inability to present a complete picture of any given situation.

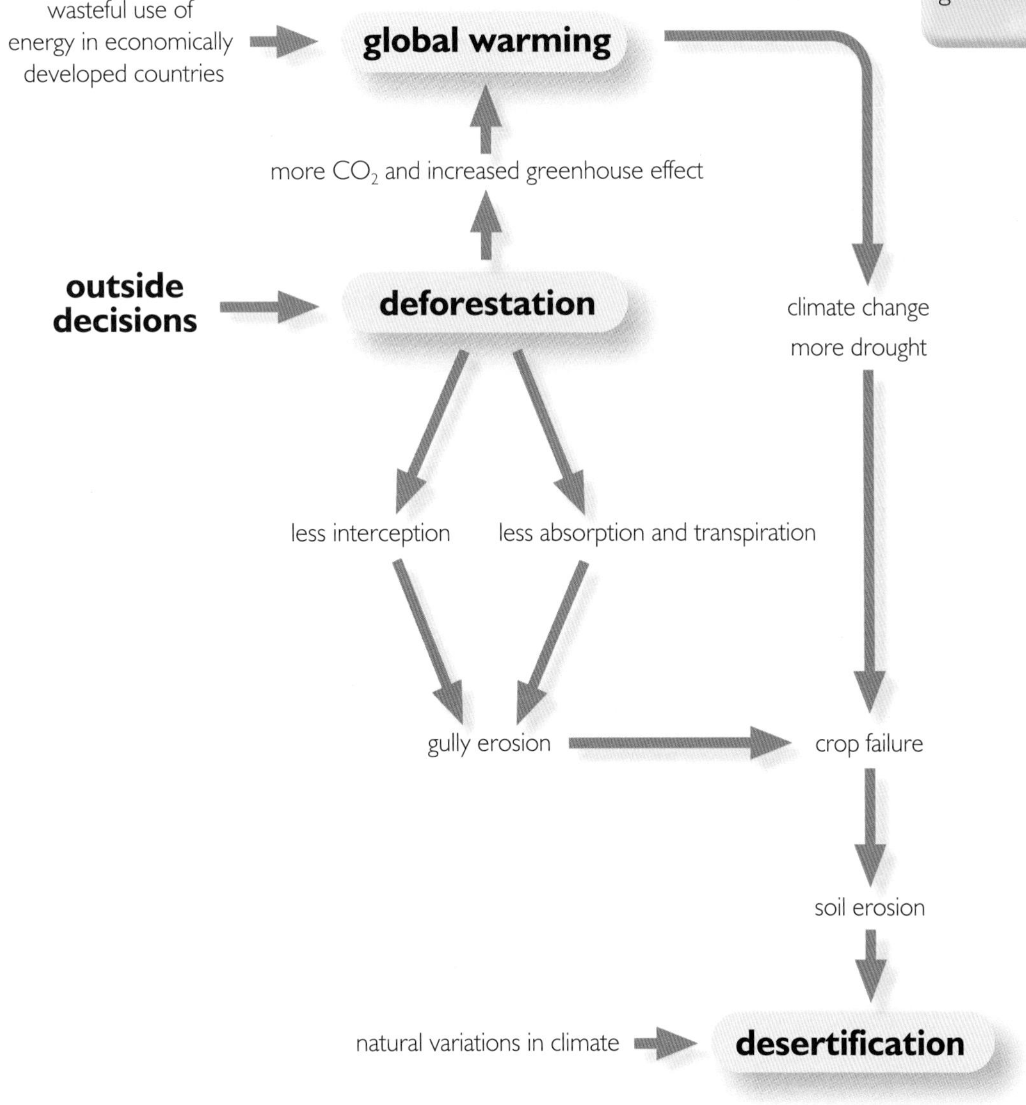

wasteful use of energy in economically developed countries → **global warming**

more CO_2 and increased greenhouse effect

outside decisions → **deforestation**

climate change
more drought

less interception   less absorption and transpiration

gully erosion → crop failure

soil erosion

natural variations in climate → **desertification**

Fig. 5.1 is a **geographical model**, which shows some of the **connections between deforestation** in the Amazon Basin, **global warming** and **desertification**. With the aid of Fig. 5.1, describe and discuss the connections mentioned.

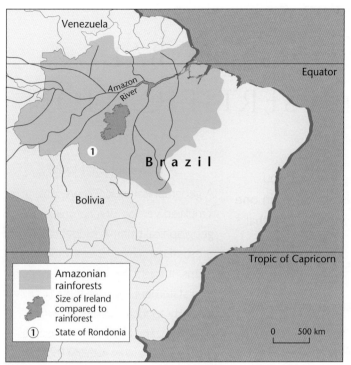

Fig. 5.2 The extent of the Amazon rainforest in relation to Ireland.

Case Study: Amazonian Deforestation

Most of the great Amazon River basin consists of a vast region of equatorial rainforest or **selva**, situated in northwest Brazil (see Figure 5.2). This selva makes up one-third of all the planet's tropical rainforests. For many centuries, the selva has been lightly populated by various groups of Amerindian peoples who did little to upset the delicate balance of the selva's ecosystems.

From the sixteenth century, a trickle of European people began to infiltrate the edges of the forest. Since the 1960s, this trickle has become an ever-increasing flow that has penetrated into the very heart of the Amazon Basin. This human flow now brings with it massive **deforestation** at an estimated rate of one hectare every one and a half seconds.

Outsiders and Deforestation in the Amazon Basin

- The general cause of the rapid deforestation of the Amazonian forests is **a concept of development** which tends to see the selva not as a unique and invaluable natural region, but rather as a 'wilderness' to be conquered in order to generate material profit. This view of development is reflected in 'The Amazon Basin's Vast Reserves' text box. It is shared by successive Brazilian governments, as well as by profit-hungry domestic and multinational logging, mining and other companies.

- **Land ownership patterns** throughout Brazil have been extremely uneven and unjust, with just two per cent of the population owning more than half of the country's agricultural land. This gave rise to demands for land redistribution by millions of land-hungry peasants in the south and east of the country. The Brazilian government, anxious to avoid land reforms which would offend the interests of the powerful landowners, used the Amazon Basin as a means of 'solving' the land problem. Poor Brazilians were lured into Amazonia with promises of free land and more prosperous futures.

> ### The Amazon Basin's Vast Reserves
> At the heart of the struggle for the Amazon are some of nature's most dazzling unclaimed prizes. Beneath the dense tropical growth of the eastern Amazon lies enough high-grade iron ore to meet world demand for four centuries. The jungle covers immense concentrations of bauxite, the reddish ore that is refined to make aluminium. Vast reserves of gold, nickel, copper, tin and timber sit untapped in the rainforest ...
>
> (*Newsweek*, 25 January 1982)

DEFORESTATION, GLOBAL WARMING AND DESERTIFICATION

- **Cattle ranching** is one of the most important and one of the most environmentally damaging land uses within Amazonia. Some large ranches, set up with the aid of government subsidies, devoted themselves to supplying American fast-food outlets with cut-price beef burgers. This practice is so unproductive that it takes five square metres of land to produce enough beef for one burger. Many ranchers farmed the land in an unsustainable manner so that it lost its fertility and began to degenerate into scrub land within a decade of the forest being cleared.
- **International monetary bodies** have played major roles in Amazonian deforestation. *The World Bank* has made huge investments in the region since 1981, funding enterprises such as the *Polonoroeste Project*. This huge settlement and farming project was responsible for widespread deforestation in the Amazonian state of Rondonia (see Figure 5.2).

What Deforestation Means to the Amazon

- **Forest peoples** face cultural and human extinction at the hands of Amazonian 'development'. One such people is the *Yanomami*, who live on both sides of the Venezuelan-Brazilian border (see Figure 5.2). These people lived independently for centuries by hunting, fishing and subsistence farming in a manner which did not upset the delicate balance of the forest ecosystem. Since the 1970s, however, the Yanomami have seen their territories invaded by 'developers' and their culture diluted by Western influences. 'Outside' diseases such as measles have killed up to ninety per cent of the people in some Yanomami villages. Traditional social life has been shattered, with many of the survivors being forced to adopt a squalid roadside existence as beggars and prostitutes.

Find out more about deforestation and the Amazon in *Our Dynamic World 2* Chapter 25.

The Yanomami, who for centuries lived in harmony with their environment (see photo on the left), now face cultural ruin and possibly extinction owing to the 'development' of their region by outsiders (see photo on the right).

Identify other minorities in various parts of the world whose culture or existence has or is being threatened by outside influences.

- It is estimated that the world's rainforests contain two million species of **plants and animals**. One-quarter of all the existing pharmaceutical drugs already owe their origins to rainforest species, despite the fact that only about one-tenth of these species have been studied by people from outside the forests. But rainforests survive in a *finely balanced ecosystem*. When this ecosystem is upset by deforestation, a multitude of plant and animal species become extinct and the land can rapidly become a barren wasteland. It is calculated that in the Amazon Basin alone, one or more species is now becoming extinct every day. An outline of the Amazonian ecosystem and the consequences of its destruction are outlined in Figure 5.3.

The ecosystem ...

(1) Trees **intercept** precipitation and so protect the soil from heavy tropical rain.

(2) The tree canopy provides **shade** for animals and small plants.

(3) Soil nutrients and water are taken up by the tree roots and gradually released through the leaves by the process of **transpiration**. This reduces the amount of water in the ground and helps to regulate rainfall.

(4) Trees shed leaves and other plant litter, which bacteria in the soil turn into **humus**. This compensates for the soil nutrients taken up by the trees.

(5) **Leaching** (washing down of soil nutrients) **is limited**.

Destroyed ...

(1) Heavy rain beats on the soil, causing **erosion**. The upper soil layers, which contain the most nutrients, are the first to be washed away.

(2) With less shade, the hot sun can harden the soil and make it less permeable. This increases **gully erosion** by surface water.

(3) More ground water means swollen rivers and **flooding**.

(4) Reduced plant litter causes the soil to become barren and eventually **desertified**.

(5) **Heavy leaching** washes soil nutrients below the reach of surviving plant roots.

Fig. 5.3 A rainforest ecosystem and some of the consequences of its destruction.

Worldwide Impacts of Deforestation

1. Global Warming and Climate Change

   The much-feared process of global warming appears to be largely caused by the gradual build-up of carbon dioxide in our atmosphere. The wasteful burning of fossil fuels within the First World is the main cause of global warming, but trees use up carbon dioxide and convert it into oxygen, which is why the Amazonian rainforests have been described as 'the lungs of the world'. Deforestation in places such as the Amazon Basin therefore plays a secondary role in global warming and in its associated climate changes which now threaten our planet. (A more detailed account of global warming is given on pages 33–36.)

2. Desertification

Deforestation and unsustainable farming in the Amazon Basin are destroying the delicate balance of the region's ecosystem and are leaving some of the land infertile and exposed to the elements (see Figure 5.3). Erosion by water can then remove topsoils and leave parts of Amazonia barren and desertified. Deforestation and associated global warming each contribute to the spread of desert conditions in many parts of the world. (A more detailed investigation of desertification is given on pages 37–39.)

GLOBAL WARMING

The world is gradually growing warmer. Glaciers are receding in the Alps and Andes, while the permafrost of Alaska and the Canadian Arctic is slowly getting thinner. Arctic sea ice has shrunk two per cent in the past twenty-five years. The 1990s were the warmest decade, and 1999 was the warmest year of the twentieth century. Most scientists agree that through the creation of what is called the greenhouse effect, humans are at least partly responsible for this global warming.

Find out more about global warming in Our Dynamic World 2 *Chapter 22 and in* Our Dynamic World 3 *Chapter 21.*

How the Greenhouse Effect Works

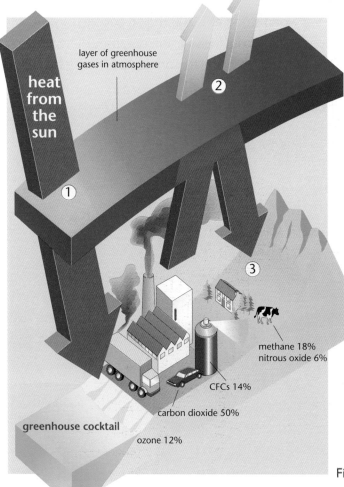

1. The sun heats the earth with **solar radiation** (visible light). This is *short-wave* radiation and **passes in** easily through the earth's atmosphere.
2. The earth converts solar radiation into **infrared heat radiation**, which it sends back into space. But some of this *long-wave* radiation **cannot easily pass out** through the greenhouse gases, which make up about one-thousandth of our atmosphere. The principal greenhouse gas is carbon dioxide (CO_2), while others include ozone (O_3) and methane (CH_4). They are called 'greenhouse gases' because they hold in the earth's heat radiation, rather like glass retains warmth in a greenhouse.
3. The greenhouse gases are essential for human survival. Without them, our planet would be a frozen wasteland with temperatures of about 30°C colder than at present. The problem is that the **greenhouse effect has been steadily growing** in recent times. Carbon dioxide and methane, which are responsible for most greenhouse activity, are each growing by about one per cent per year. This growth is causing our planet to gradually become warmer and is likely to have severe human and ecological consequences.

Fig. 5.4 shows how the greenhouse effect works.

Some Causes of Global Warming

The principal human cause of global warming is the **burning of fossil fuels** such as wood, coal, oil and gas. When these fuels are burned, carbon dioxide is released into the air and increases the greenhouse effect.

- **Economically developed countries** are overwhelmingly responsible for the excessive burning of fossil fuels. Industrialisation, increased vehicle traffic, intensive livestock farming and the wasteful use of domestic energy within the First World have all hugely contributed to global fossil fuel usage. The five per cent of the world's population that lives in the United States emits almost a quarter of the global emissions of CO_2. India, on the other hand, contains sixteen per cent of the human family but produces only three per cent of global CO_2 emissions.

- **Deforestation**, especially in regions such as the Amazon Basin, contributes significantly to global warming. Forests – especially dense rainforests – are referred to as *carbon sinks* because they absorb carbon dioxide and so help to offset the effects of global CO_2 emissions. The destruction of the forests not only destroys potential carbon sinks, but also leads to *the release of carbon into the atmosphere*, which is stored in wood. The sheer scale of global deforestation gives rise for serious concern. It is estimated that rainforests twice the land area of Ireland are being destroyed throughout the world each year. This threatens the entire world with a significantly increased greenhouse effect.

- **Rapid population growth** has also indirectly contributed to global warming. In 1850, the total population of the world stood at 1,200 million people. Now it exceeds 6,000 million. Most people use fossil fuels for the purposes of heating and cooking. Therefore, population growth has indirectly contributed to an increase in the greenhouse effect.

Clearing the rainforest.
(a) Name three regions in the world in which this scene might take place.
(b) Describe the process shown and explain fully how this process could contribute to global warming and desertification.

Some Consequences of Global Warming

If greenhouse gases continue to accumulate at their present rate, it is predicted that the earth's temperature will rise by possibly 3°C in the next one hundred years. This could have several dramatic effects on our planet:

- The melting of polar ice caps, which appears to have begun already, will accelerate. It is estimated that because of this, **sea levels will rise** by at least 50 cm during the present century. This could result in the *disastrous flooding* of several coastal and low-lying regions of the world. Figure 5.5 shows the principal regions that may be threatened by sea flooding. They include parts of the Netherlands, which is the world's most densely populated country. Irish areas likely to suffer from flooding would include the lowlands of the Shannon Estuary, where more than 19,000 hectares could be at risk within the next forty years. In Dublin Bay, areas around Sandymount and Sandy Cove are also considered to be at some long-term risk.

Fig. 5.5 Some areas threatened by flooding if the sea level rises. Identify six major regions shown here which might be flooded as a result of global warming.

- Global warming is likely to cause profound **agricultural changes** in many Third World regions. African countries such as Morocco, Tunisia, Mali and Nigeria would suffer greatly from increased drought, crop failures and *desertification*. This would give rise to poverty and hunger, which would in turn result in massive *human migrations* to other parts of Africa.

    A positive result of rising global temperatures might be longer growing seasons and *increased agricultural production in temperate countries* such as Ireland. Grass production in Ireland could increase by twenty per cent. It is possible that we will yet be able to grow crops such as vines, sunflowers and peaches on a commercial basis.

    This advantage might be countered, however, by more *cloud* and prolonged heavy *rains* in winter as well as by periods of *drought* and water shortage in summer. Oceanographers also fear that the melting of ice caps might result in the *North Atlantic Drift* taking a more southerly course than at present. If this happened, ocean temperatures might decrease to the extent that our harbours would freeze in winter, with disastrous results for our fishing industry. Prevailing onshore winds from colder seas might then result in a decrease in atmospheric temperatures over Ireland.

- It is feared that serious tropical **diseases** might spread to temperate countries such as Ireland. *Malaria* is at present one of the biggest killers on the planet. With global warming, by 2050 malaria-bearing mosquitoes could easily spread to temperate countries such as those of Western Europe. The UK Department of Health has already instructed doctors to look out for possible outbreaks of malaria in southern England.

  Global warming will probably result in more instances of sunburn and therefore greater risks of **skin cancers** within Europe. Irish people, who are among the most pale-skinned in the world, would be particularly vulnerable to the ravages of skin cancer.

- **World tourism** could be radically altered by continued global warming. While warmer summers in *Ireland* might entice large numbers of foreign tourists to our shores, many existing *Mediterranean tourist resorts* are likely to suffer serious decline. It is estimated that by the year 2050, Greek islands such as Crete could be too hot for tourism during the peak tourist months of July and August.

  Winter holidays are likely to be badly affected by global warming. High-altitude ski resorts such as Zermatt in Switzerland will probably survive, but less elevated Swiss resorts such as Splugen may no longer have the snow levels to sustain skiing. If resorts become fewer in number, the costs of winter tourism can be expected to rise correspondingly.

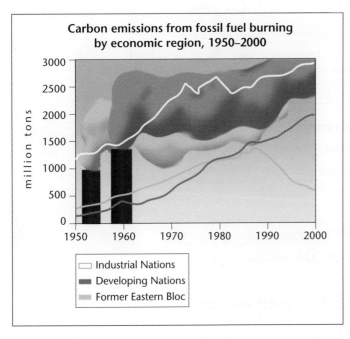

Fig. 5.6 Carbon emissions by economic region.

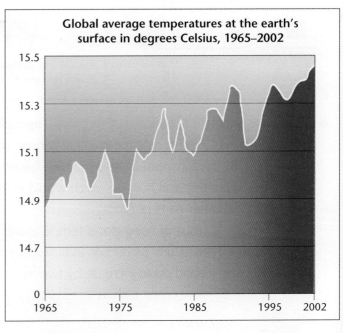

Fig. 5.7 Global average temperatures at the earth's surface.

Describe, account for and examine the consequences of the trends revealed in Figure 5.6 and Figure 5.7.

DESERTIFICATION

Our world's deserts are growing, *spreading gradually into nearby semi-arid areas*. This process is referred to as **desertification** and is one of the great environmental and human disasters of our times.

Desertification now threatens the lives and livelihoods of hundreds of millions of people. It occurs to some degree in virtually every semi-arid region of the world and is beginning to happen even in those parts of the Amazon Basin that have suffered from inappropriate development. But the area that is the worst affected by desertification is the southern edge of the Sahara Desert. This vast region, which stretches for almost six thousand kilometres east to west across Africa, is known as **the Sahel** (see Figure 5.8).

> Find out more about the Sahel in *Our Dynamic World 1* page 145 and in *Our Dynamic World 2* Chapter 25.

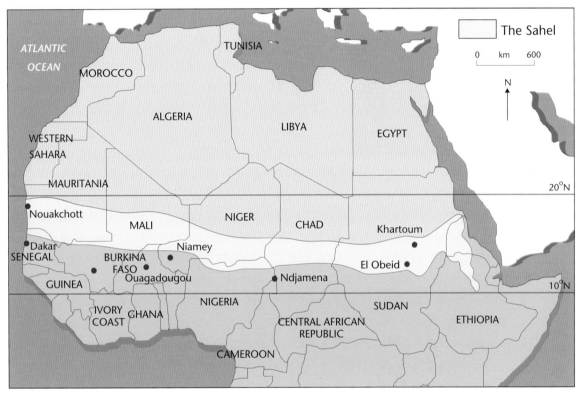

Fig. 5.8 The Sahel
(a) Use the information on the map to describe the extent and location of the Sahel. Refer to the approximate length, width and area of the region. Refer to its latitude, its orientation (the direction in which its long axis points) and the countries which it partly occupies.
(b) 'Sahel' is the Arabic word for 'shore'. Why do you think the Sahel is so called?

Scientists believe that three factors work together to cause desertification in the Sahel:
- The precipitation levels of this semi-arid region have a **natural** tendency to fluctuate over long periods. Since the 1960s, a period of higher-than-usual rainfall has been replaced by periods of prolonged drought.
- Human actions in other parts of the world are contributing to **global warming**, which contributes to climate change.
- People have overcultivated and overgrazed the Sahel and have deforested areas to procure firewood. These activities result partly from local population growth, and partly from actions and decisions taken far outside the Sahel.

Case Study:
Peanuts and Desertification in Niger

The Sahelian country of Niger (see Figure 5.8) was a colony of France until 1960. Like all imperial powers, France encouraged its colonies to grow **cash crops** to supply French consumers with cheap produce which could not be grown at home. One of the principal cash crops produced in Niger was **groundnuts** (peanuts).

In the 1960s, **decisions** were being made in the *United States* and *France* which were to contribute to disastrous desertification in Niger. The United States at this time decided to boost its economy by exporting large quantities of cooking oil and other soya bean products to Europe. The French government responded to this competition by seeking alternative sources of cooking oil for itself. It decided that groundnuts, grown in the former colony of Niger, would provide the answer.

The French government and private French companies encouraged Niger to grow more groundnuts for export. As a result, the **groundnut growing area** of that colony **increased** more than three-fold between 1954 and 1968.

Throughout the 1960s, Niger became part of a global agricultural economy. Farmers there became so reliant on groundnut exports for their incomes that food production fell and food imports became necessary. The following **problems** then began to arise:

- The *terms of international trade* worked against Third World countries such as Niger. While these countries had to meet the ever-rising costs of First World imports, the prices they received for 'unprocessed' products such as groundnuts fluctuated wildly on the world markets. In 1965, the price of groundnuts actually began to fall. Within three years, prices had fallen by approximately twenty-five per cent.
- The monocultural (exclusive) growth of groundnuts so exhausts the soil that following three years of its cultivation, the land needs to be left fallow (rested) for a six-year period. Because most farmers could not afford to rest their land for such long periods, the *soil gradually became infertile*.
- The French government tried to combat infertility by exporting fertilisers to Niger. This remedy was not appropriate because Niger's farmers simply could not afford to buy fertilisers. Many farmers had already gone deeply into *debt* to purchase the seed and technology needed to produce groundnuts on a large scale.

In order to maintain living standards, Niger's farmers felt they had no option but to **further increase** groundnut production. To do this, they began to use fields that should have been left fallow. As more land was devoted to groundnuts, traditional millet cultivators and cattle herders were pushed into 'marginal', semi-desert areas. During particularly dry years, the poor, marginal land would bear little or no crops, leaving the soil dry and exposed to wind storms. As the wind carried the dusty topsoil away, the land became part of the encroaching Sahara desert. **Desertification** had taken place.

Consequences of Desertification

Desertification has greatly **increased human poverty** and suffering in countries such as Niger:

- Crops have failed and many thousands of cattle have died of **hunger**. People have been left without adequate incomes or food supplies. This has caused an increase in malnutrition, especially among the very young. Many children of the Sahel die of **diseases** associated with malnutrition, such as Kwashiorkor, a condition which results in thin, wasted limbs, swollen stomachs and an almost complete lack of energy among its victims.

- Many people in the worst-affected areas have no option but to move away in order to survive. Hundreds of thousands of people have **migrated** southwards, contributing to overpopulation and a further spiral of land overuse and desertification in the lands to which they moved. Others have left for cities such as Niamey in Niger or Ouagadougou in Burkina Faso (see Figure 5.8). As these cities became overpopulated, the demand for firewood among their inhabitants has caused **deforestation** of the limited numbers of trees that existed in their urban hinterlands. This, in turn, contributes further to the spiral of **land overuse and continued desertification**.

Comment on the possible cause and effects of **Kwashiorkor** on this child.

Activities

1. Define each of the following terms:
   - deforestation
   - global warming
   - desertification
   - geographical model.
2. With reference to one deforested region that you have studied, comment on each of the following statements:
   'Deforestation in a region is often caused by people outside of that region.'
   'Large-scale deforestation can have serious global effects.'
3. Describe the effects that large-scale deforestation can have on a region's ecosystem.
4. (a) Explain how global warming takes place.
   (b) Outline the links between deforestation, global warming and desertification.
5. With reference to any desertified country or region that you have studied, explain how outside decisions or actions can give rise to desertification, which in turn can have adverse effects in other countries or regions.
6. Explain why you agree or disagree with each of the following statements:
   'The developed world is to blame for the threat of climate change.'
   'Climate change could bring several benefits to countries such as Ireland.'
   'Today's global economy benefits the environmental welfare of our planet.'

CHAPTER 6
THE MOVERS

In the two previous chapters you learned that in an interdependent world, manufacturing or environmental decisions taken in one area can impact on other areas. This chapter will explain how **social or political decisions taken in one area can set up human migration patterns which may impact on other areas**.

Find out more about changing migration patterns in Our Dynamic World 3 *Chapter 6.*

Some Useful Terms Relating to Human Migration

- **Migration:** The movement of people over a considerable period of time for the purpose of living in a different administrative unit. The term 'different administrative unit' can refer to a different country, state, city or other region. A 'considerable period of time' is frequently, though not always, taken to mean one year or more.

- **Internal migrant:** A migrant who moves within a country.

- **International migrant:** A person who moves from one country to another. Such a person would be an **emigrant** from the country from which he/she leaves, and an **immigrant** in the country into which he/she enters.

- An **economic migrant:** One who migrates for economic reasons, such as to improve one's standard of living or to escape poverty.

- A **refugee:** Someone who, owing to a well-founded fear of persecution for reasons of race, religion, nationality or membership of a particular social group or political opinion, is outside his/her country of nationality and is unable or, owing to such fear, is unwilling to avail of the protection of that country. (Definition of the Geneva Convention, 1951.)

- An **asylum seeker:** Someone who, on grounds of being a refugee, formally requests permission to live in another state and is in the process of having his/her case examined by the government of that state.

- An **illegal immigrant:** Someone who has entered a country without permission and whose presence in that country is therefore undocumented.

- **Push factors:** Circumstances that encourage people to *leave* their areas of origin. Lack of employment is the most common push factor. Other push factors include war, famine, persecution and social or environmental problems.

- **Pull factors:** Circumstances that encourage people to move *into* an area. The prospect of economic improvement is the most common pull factor. Other factors could be freedom from oppression or the perception of better social or environmental conditions in the destination area.

- **Racism:** The idea that some ethnic groups are naturally superior to others, implying that those of a 'superior' group might be entitled to dominate or otherwise abuse those of an 'inferior' group.

SOCIAL AND POLITICAL DECISIONS RESULTING IN HUMAN MIGRATION

Social Decisions and Economic Refugees

1. Rural-to-Urban Migration in the Third World

Colonial governments in the past and powerful international bodies such as the World Trade Organisation (WTO) of today have used their influence to increasingly **'Westernise' and globalise world agriculture**. More and more agricultural production is being controlled by relatively few powerful landowners or by huge 'agribusiness' companies rather than by numerous peasant smallholders. This trend has contributed to increasing mechanisation, landlessness and unemployment in rural areas. Rural unemployment has been a major factor in **rural-to-urban migration**, which is the greatest single form of human migration in the world today.

Rural-to-urban migration has taken place in almost all countries, and includes the movement of people from the west of Ireland to Dublin and from France's Massif Central to Paris. However, this form of economic migration is most common in Third World countries, where most of the world's country people live, where agriculture dominates employment and where the populations of large cities are swelling rapidly. Rural-to-urban migration has now resulted in so much urban growth in the South that more than two-thirds of all the world's city dwellers now live in developing countries. Third World cities that have experienced phenomenal population growth include Mexico City, Sao Paulo (Brazil) and Calcutta (India).

City	1990 Population (Millions)	2000 Population (Millions)
Mexico City, Mexico	19.4	24.4
Sao Paulo, Brazil	18.4	23.6
Tokyo, Japan	20.5	21.3
New York City, USA	15.7	16.1
Calcutta, India	11.8	15.9
Bombay, India	11.1	15.4
Shanghai, China	12.6	14.7
Tehran, Iran	9.2	13.7
Jakarta, Indonesia	9.4	13.2
Buenos Aires, Argentina	11.6	13.1

Table 6.1 Growth of the world's ten largest cities, 1990–2000.
(a) Which of the world's ten largest cities are in the First World?
(b) Rank the ten cities in Table 6.1 according to their population increases between 1990 and 2000. Comment on the positions of First World cities within this ranking.

Pavement people in Calcutta.
The population of Calcutta has now grown to sixteen million people. Explain how decisions taken in other areas have contributed to the growth of Calcutta. How has rapid urban growth affected living standards in Calcutta? Why are some of the people shown here called 'pavement people'?

Find out more about Calcutta in *Our Dynamic World 1* page 258 and in *Our Dynamic World 3* pages 150–51.

Revise 'Peanuts and Desertification in Niger' on pages 38.

2. Ecological Refugees in Niger

In response to commercial competition from the United States, in the 1960s the French government decided to encourage the commercial growth of **groundnuts** in the former French colony of Niger. Increased groundnut production soon became a vital part of Niger's agricultural economy, but the intensive growth of this cash crop in the absence of suitable but expensive fertilisation gradually damaged soil fertility. As infertility set in, crops began to fail and the exposed soil was damaged by wind erosion. Such erosion contributed to **desertification** in the area, which in turn resulted in the mass out-migration of country people from their ancestral lands. Thousands migrated southwards into other rural areas, contributing to overcultivation, overgrazing and the consequent desertification of these areas. More economic refugees fled to **Niamey**, the capital of Niger, contributing to rapid population growth in the city.

Political Actions and Political Refugees

1. The Creation and Collapse of Yugoslavia

Throughout the 1990s, more than five million people fled from Bosnia, Kosovo and other regions of what was formerly the state of Yugoslavia. These were mainly **political refugees** who were fleeing discrimination, persecution and armed conflict in the region. Some had been victims of 'ethnic cleansing', whereby minority groups were driven from their homes by hostile majorities. Most refugees migrated within the territory of former Yugoslavia, while others fled to neighbouring European countries such as Italy. Over one thousand Bosnians found their way to Ireland, where they sought asylum.

The eruption of conflict and population movements in the former Yugoslavia resulted partly from a series of **political decisions and events that took place outside the Balkan region** in which Yugoslavia was situated:

- The state of Yugoslavia was set up in 1919 at the urgings of France, Britain and the United States, who were the chief allied victors in World War One. The new state was established largely to deprive the defeated Austrian empire of its territories in the region known as the Balkans. But Yugoslavia was essentially **an artificial state**. It contained many different ethnic groups who were deeply suspicious of each other. Such groups included Orthodox Christian Serbs, Catholic Croatians and Muslim Bosnians.

- In 1941, **Nazi Germany decided to invade Yugoslavia** and, with the help of some Croats, set up a puppet state in northern Yugoslavia. Communist partisans and Yugoslavian nationalists resisted Germany. In the four years that followed, one-tenth of Yugoslavia's population was killed, mostly in fighting between fellow Yugoslavians. At the end of World War Two, Yugoslavia was reunited and civil strife was halted under the communist leader Marshal Tito. But the events following the German invasion created *lasting enmity* between different sections of Yugoslavia's population.

- With **the collapse of Eastern European Communism** in 1990, the last thin threads holding Yugoslavia together were broken. Almost immediately, the country rapidly disintegrated into small and often mutually hostile states (see the map). The conflict and persecutions that followed resulted in widespread, forced migrations from Bosnia, Serbia, Croatia and other new states.

Fig. 6.1 The former Yugoslavia.
Identify the states that make up the former Yugoslavia. Why might Yugoslavia be described as 'an artificial state'?

2. Northern Ireland

In 1920, the British government made a decision that was to contribute to political migrations within Ireland. In that year, the **Government of Ireland Act** split the island of Ireland in two and set up a parliament at Stormont, Belfast to govern the counties of Down, Antrim, Armagh, Fermanagh, Tyrone and Derry. While Northern Ireland, as the new political entity was called, was to remain part of the United Kingdom, the Stormont Parliament was given power over the internal running of the region.

Northern Ireland was dominated by an almost two-thirds Unionist majority. Unionists were mostly Protestant and wished to remain within the United Kingdom. A significant minority, however, were Nationalists. They were mostly Catholics who wished to become part of an independent Irish Republic.

Unionists and Nationalists led lives that were culturally different and many were actively hostile to each other's political views and cultures. Political divisions spawned **discrimination**, repression, terrorism and protracted periods of **violence**, which led in turn to political migrations within and out of Northern Ireland. Some Catholics were forced to abandon their homes in mainly Unionist areas (see extract below). Protestant minorities were also forced out of some mainly Nationalist housing estates. From 1969 to 1972, when the 'Troubles' in Northern Ireland were at their worst, many thousands of Nationalist political refugees fled southwards into the Republic.

> Find out more about Northern Ireland in *Our Dynamic World 1* pages 270–73.

> Irene and her family received the death notice to move out of their north Belfast home exactly twenty-six days after the first attack had happened.
>
> The following morning Irene, her husband and their three sons left the home they had lived in for fourteen years in what was once a quiet and religiously mixed estate off the Antrim Road, but has in recent years seen Catholics driven out.
>
> (*The Irish Times*, 2 July 1999)

Repression in Belfast: Catholic children on their way to Holy Cross primary school are attacked by a Unionist mob.

PATTERNS OF MOVEMENT

While migratory patterns are extremely complicated some general global patterns are outlined in Table 6.2.

Patterns	Some Causes and Examples
1. Rural to urban (internal migration)	Most migrants are in search of **employment** and better **social and cultural infrastructures**, such as hospitals, colleges and sources of entertainment. Examples include migrations from *rural India into Calcutta* and from the *west of Ireland to Dublin*. A flight from **ecological disasters**, such as desertification in the Sahel region of Africa, is another cause of internal migration. An example would include *rural migration into Niamey*, the capital city of Niger. Some migrants flee from areas of **conflict**. For example, civil unrest in *Angola* caused migration into the capital city of *Luanda* in 1999.
2. International migration between Third World countries	Over eighty per cent of all international migration is between Third World countries. A principal cause of such movement is the search for **employment**, as from *Mozambique to neighbouring South Africa*. Flights from **ecological disasters** also cause such migration, as from the *Sahel region* in Niger into nearby Nigeria. Other people migrate to neighbouring countries to avoid **war or persecution**, as in the case of the flight of refugees from *Rwanda into the Democratic Republic of Congo* in 1994.
3. International migration from Third World to First World countries	The main migratory trends under this category are from *Central and South America to North America and from Africa, the Middle East and China to the EU and the USA*. **Poverty and unemployment** in these source regions, coupled with demographic stagnation and the consequent **demand for labour** in the EU and USA, have been the principal causes of this migratory trend.
4. Migration from Eastern to Western Europe	The collapse of Communism in Eastern Europe in the early 1990s has been followed by social and economic upheavals in much of the region. This has contributed to large-scale migration to Germany and other Western European countries. Most migrants move for **economic** reasons, but some gypsies have been forced to flee from **discrimination**. It is estimated that between 2001 and 2006, two million people will migrate from East to West.
5. Migration from peripheral to core regions within the EU	Searches for **employment and higher living standards** have been the principal causes of this migratory pattern, while the desire for better **social and cultural infrastructures**, ranging from education to nightlife, is a contributory cause. Examples include the movement of people from *southern to northern Italy* and from *northern England to the London area*.

Table 6.2

Immigration Trends and Ireland

From the Great Famine up to the mid-1990s, Ireland had been a source of rather than a destination for most international migrants. For 150 years, our country was one of the world's most persistent suppliers of economic migrants, who left our shores to seek employment in England, the United States, Canada and other countries.

This situation changed dramatically in the mid-1990s, when Ireland's '**Celtic Tiger**' economy began to boom and provide plentiful employment within the Republic. The flow of economic migration was reversed, as greater numbers entered rather than left the country in search of employment. Many **immigrants** have been returning Irish people who had once emigrated. Most others have come from the United Kingdom, France, Italy and other EU member-states.

Among the immigrants to our shores are small numbers of **refugees** from outside the EU. Some of those are economic migrants who, like our Irish ancestors, have fled from extreme poverty in their home countries. Others are political refugees who flee from war or persecution and who seek asylum within our land. In 2000, Ireland received just under 11,000 applications for asylum. However, only 200 people were granted refugee status in that year – one for every 23,500 of our population.

Year	Number Entering Ireland	Granted Refugee Status	Granted Humanitarian (Temporary) Leave to Remain
1997	3,883	211	5
1998	4,626	88	0
1999	7,724	166	40
2000	10,938	200	19

Table 6.3 Asylum seekers in Ireland.
(a) Describe the trends of asylum seekers entering Ireland between 1997 and 2000.
(b) Comment on the proportions of asylum seekers who have been granted refugee status or humanitarian leave to remain.

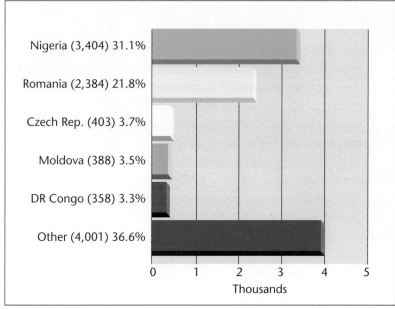

Fig. 6.2 Main countries of origin of asylum seekers in Ireland, 2000.
(a) In the case of one country identified, suggest why it might be a source of refugees to Ireland.
(b) Name one country that you think might come under the category of 'other', and outline precisely why you think this country would be a source of refugees.

IMMIGRATION AND HUMAN RIGHTS ISSUES

Immigrants are often vulnerable to racial discrimination and other forms of abuse. Examples of such abuses of human rights are common in many parts of the world:

- It has been estimated that in 2000 alone, approximately 125,000 economically poor young women and girls were trafficked illegally from Russia, Ukraine, former Yugoslavia and other Eastern European countries for the purposes of **sexual exploitation** in Western Europe and America. Many of these women are forced to work under conditions of virtual slavery. If discovered by Western authorities, the women often tend to be treated as criminals while relatively little focus is given to apprehending the criminal networks which traffic them.
- Italy considers itself to be a less racist country than most others in Europe, yet **racist attacks** on refugees and other immigrants are not uncommon. An Italian government source admitted in 1997 that immigrants to Italy were being murdered at a rate of one every three days.
- Britain appears to **discriminate** in favour of white people and against Africans and Asians who request settlement in the United Kingdom. In 1997, less than one in 500 Australians and less than one in 200 Americans requesting settlement in Britain were refused. For people from India and Ghana, the refusal ratio for the same year was almost one in three.

The abuse of immigrants – an old concern.
This cartoon was published in a United States newspaper in the 1920s.
Describe fully the situation being illustrated by the cartoon.

Ireland, Refugees and Human Rights

Find out more about Irish attitudes to refugees in Our Dynamic World 3 pages 51–2.

Up to 1997, Ireland was a source rather than a destination country of international migrants. The few immigrants who did reach our shores tended to be met generally with an attitude of friendly curiosity. From 1997, as the number of immigrants to Ireland began to increase rapidly, so did reports of immigrants being abused by a minority of Irish people.

The Case of Mr Luyindula

Mr Luyindula is a journalist from Zaire who has sought asylum in Ireland. He escaped from Zaire having been imprisoned for criticising its then dictator, Mr Mobutu.

Life in Ireland has not been a happy experience for Mr Luyindula. He was beaten up by a group of thugs in Temple Bar in Dublin. Nobody stopped to help him. An anonymous racist letter was then sent to 'The African' at his flat, and he has been told to 'go back to Africa' several times by people in the street.

'When I came here everyone seemed friendly,' said Mr Luyindula, 'but since the attack I have started to see some Irish people in a new light. Many of them think it's all right to have one or two black people living in Ireland. However, more than that irritates them. Older people are usually nice, but younger people are sometimes nasty and become particularly aggressive when they are drunk.'

(*Human Rights and Refugees*, Trócaire)

Discuss Mr Luyindula's views on Irish people. Do you think they are well founded?

Racism and Discrimination

- A survey in 1999 by a Catholic organisation called the Pilgrim House Community found that ninety-five per cent of African asylum seekers to Ireland had suffered **verbal abuse**, mostly on a daily basis. More than one in five had been **physically assaulted**. Some asylum seekers in Dublin have received racist hate-mail (see 'The Case of Mr Luyindula' text box).
- It has occasionally been assumed by some that anti-immigrant attitudes tend to be more prevalent among people who are economically deprived and poorly educated. Yet a survey among **Irish third-level students** in 2001 showed that one in five Irish third-level students felt that illegal immigrants should be sent home to their countries of origin without exception.
- Some immigrants feel that they are discriminated against even by authority or **establishment figures** in Ireland. African and Romanian refugees suffer from a widespread impression that they are viewed by Irish authorities as being potential criminals. In 1997, when the numbers of asylum seekers to Ireland began to rise, sections of the media went on a spree of alarmist reporting about the state being 'swamped' by 'bogus' refugees and other undesirables.

The Irish Government and Refugees

There have been signs that the Irish government has sometimes tended to react to immigration in an alarmist and even discriminatory manner. In 2000, the Taoiseach hinted that illegal immigrants might be imprisoned, while our Minister for Justice at that time appeared to favour the accommodation of asylum seekers in 'floating hotels' – described as 'prison ships' by his critics – off the Irish coast.

Our official treatment of refugees and asylum seekers is based on the terms of the **1996 Refugees Act**. Some positive and negative aspects of this Act and its application are outlined below.

The Refugee Act

Positive Aspects
- Those who arrive in Ireland *cannot be forced to return* to their country of origin if it can be established that their lives or freedom are threatened in that country.
- Arrivals cannot be deported without having been first informed of their *right to apply for refugee status*.
- Applications for refugee status must be heard and processed fairly. An *independent Refugee Commissioner* must recommend whether or not applicants should receive refugee status.
- An applicant who is refused asylum has the *right to appeal*.
- Those who are granted refugee status are given the *same rights as Irish people* in matters relating to health care, education, religious freedom, access to employment, social welfare, etc.

Negative Aspects
- Asylum seekers receive *no free legal aid or* guaranteed right to *interpreters*. There is no provision for the special training of officials to deal with asylum seekers.
- Asylum seekers are *not permitted to work or to leave the country* while their application is being processed. They have no right to state-funded language classes, education or training.
- At present, asylum seekers may have to *wait for years* to have their applications processed.
- The Refugee Act *does not recognise* as refugees '*economic migrants*' who leave their home countries because they cannot find employment there.

Asylum seekers in Ireland. Most asylum seekers are housed in hostels, mobile homes or bed and breakfast accommodation. They are given their meals and around €19.05 a week (2003 figures) to live on. Most asylum seekers are dispersed to one of several villages or towns around Ireland. Some local people complain that inadequate support structures and services are put in place before the asylum seekers arrive. Some asylum seekers complain of feeling isolated.

Ways of Improving the Treatment of Immigrants

Education

It is important that our education system and our media should actively support **enlightened and humane principles** relating to immigration. These principles might include the following:

- The principle that all human beings are essentially part of a *single human family*. This belief carries with it the assumption that each member of the human family has an obligation to help other, less fortunate members.
- A realisation that the 'asylum seeker problem' in Ireland stems partly from Ireland's difficulty in moving from a situation of cultural insularity to a state of *interculturalism*. Instead of focusing solely on the treatment of asylum seekers, we should also focus on what can be done to assist Irish society in making this transition.
- We must accept that immigrants to Ireland possess and are entitled to *traditions* that are different from those already existing in Ireland. We should also realise that blending the best of new and existing traditions is likely to enrich rather than weaken our cultural identities as Irish people.

Assistance for Voluntary Bodies

More state support could be given to assist voluntary bodies such as the Irish Refugee Council and Trócaire, which play positive roles in Ireland's response to immigration. *The Irish Refugee Council*, with the help of unpaid volunteers, provides asylum seekers with advice on law, housing and social welfare services. Organisations such as *Trócaire* do much to educate Irish opinion on the issues associated with immigration and on the needs of asylum seekers and other immigrants.

An asylum seeker being assisted by an Irish Refugee Council volunteer. Why is such assistance important?

Interculturalism is the promotion of understanding, interaction and integration between different cultures and ethnnic groups.

Activities

1. Briefly but clearly explain the meanings of each of the following terms that relate to human migration:
   - refugee
   - asylum seeker
   - internal migrant
   - pull factor
   - racism.
2. In the case of countries or regions that you have studied, demonstrate how socio-economic or political decisions made in one area have resulted in human migration in other areas.
3. With reference to the information in Figure 6.3 describe and attempt to account for the principal migratory patterns shown.
4. (a) Comment on the extent to which the human rights of refugees and asylum seekers are being met in the Ireland of today.
   (b) Offer some suggestions on the ways in which these rights could be met more fully.

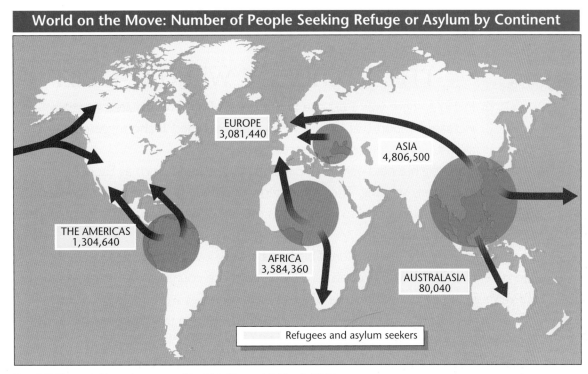

Fig. 6.3 The world on the move. *The Irish Times*, 10 March 2001.

CHAPTER 7
INTERNATIONAL DEBT AND CYCLES OF POVERTY

> **Debt Facts!**
> 1. In the Third World a child dies every three seconds because of hardships related to international debt.
> 2. Each person in the Majority World 'owes' several hundred euro to international banks. This is more than many earn in an entire year.
> 3. For every euro received by the South in aid, the South pays nine euro in debts to the North.

Many Third World governments and aid agencies work hard to empower people so that economic growth and human development can proceed together in the Majority World. Improved health, education and human rights are all vital to the empowerment of people.

Among the greatest **obstacles to human empowerment** and development are the large financial debts that developing countries owe to international private banks and to international world bodies such as the World Bank. The payment of these debts has placed an intolerable burden on Third World governments and unbearable hardships on many Third World people.

THE STORY OF WORLD DEBT

In 1973, there was a big **rise in the price of oil**. This resulted in greatly increased import bills for all oil-importing countries. It also triggered an almost worldwide *economic recession*.

Leaders and big business people of oil-producing countries such as Saudi Arabia and Kuwait became hugely wealthy. They deposited vast amounts of oil profits in **Western banks**. The banks found it hard to reinvest this money in Europe and America because of the economic recession.

In desperation, the banks offered big **loans** at very low interest rates to governments of Third World countries. The banks chose countries that were considered relatively safe places to invest. Some of these 'safe' countries were ruled by dictators who were considered 'pro-Western'. Such countries included the Philippines ruled by President Marcos, and Zaire under President Mobutu.

> Find out more about borrowing in *Our Dynamic World 2* Chapter 6.

Borrowed money was used wisely by some Third World governments to help develop their countries and empower their peoples. Tanzania, for example, invested in health and education programmes. But in other countries **money was wasted** on weapons or on useless 'prestige' schemes or was simply stolen by dictators. Mobutu of Zaire embezzled enough cash to purchase several mansions in Europe. Marcos invested in a nuclear plant on the site of a volcano! (The plant was never used, but cost $2.2 billion of borrowed money.) A dictator in the Central African Republic decided to finance a cathedral larger than St Peter's in Rome. The bankers did not care much, as long as their Third World investments appeared to be secure.

On the surface, at least, all went well until the early 1980s. Then disaster struck for the following reasons.

The United States had been spending far more than its income, particularly on military equipment. It therefore needed to attract capital from abroad. To do this, it raised its interest rates sharply. When the United States increased its interest rates, other Western countries did likewise. **World interest rates soared**.

Rising interest rates made it virtually impossible for poor Third World countries to manage their debts. The national **debts** of many Third World countries grew quickly and soon went **out of control**. In 1982 Mexico declared that it simply could not repay its debts. By the mid-1980s, many people began to argue that the ordinary people of the Majority World should not be obliged to repay debts given by foolish bankers to unelected dictators.

Western banks and governments fell into panic, fearing that nonpayment of Third World debts might result in the collapse of the banking systems of the West. They clubbed together with the **International Monetary Fund** (IMF) to ensure that international debts would be 'rescheduled' rather than not paid.

The banks agreed to delay repayments, provided that Third World debtor countries agreed to IMF **structural adjustment programmes**. These programmes forced debtor countries to 'restructure' their economies so that they could earn more money from exports and so more easily pay their debts. The programmes typically entailed the following:

- Increased production of *cash crops for export*: This usually resulted in less land being used to produce food crops for local people. Food shortages became more common.
- *Reduction in 'unprofitable' public spending*: This meant less money for schools and health care for poor people.
- *Abolition of price control* so that prices could rise or fall to 'market values': This meant that governments such as that of Zimbabwe could no longer subsidise the price of maize, which is the staple diet of poor people. The price of such commodities rose dramatically.
- *Free movement of capital out of Third World countries*: This enabled foreign transnational companies and investors to move profits more easily, but it resulted in 'capital flight' from Third World countries and reduced the control of Third World governments over the economies of their countries.

Poor quality school in Mali, Africa. Explain how international debt contributes to poverty such as this.

- *Devaluing of Third World national currencies:* This favours exporting by making exports cheaper, but imports became more expensive and this caused the cost of living to rise sharply.
- The implementation of *wage restraint and wage freezes*: This reduced inflation but made economic survival more difficult for working people. Civil unrest followed in countries such as Zambia and Zimbabwe.

Structural adjustment programmes helped to protect international banks from the threat of unpaid international debts, but they also resulted in falling living standards for the poor of the Third World and in an even wider gap between the rich and the poor of our planet.

International Debt and Cycles of Poverty

The people of the Majority World were already poor when their countries first went into debt with Western banks. The payments of these debts and structural adjustment programmes have contributed to keeping the people poor. **International debts have therefore contributed to the cycle of poverty, which is endured by most people in the Third World (see Figure 7.1).**

Fig. 7.1 Debt and the cycle of Third World poverty.

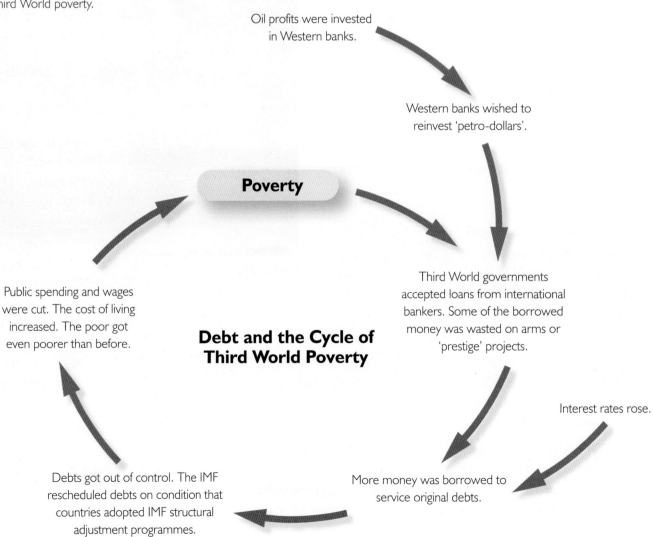

What Should Be Done?

International debt is now recognised as a major problem facing humanity. Some moves have already been made to cancel some of the debts owed by Third World countries. By January 2000, the world's richest countries had agreed to cancel about thirty per cent of the total amount owed by the world's poorest countries. But more needs to be done if the pressure of debt is to be lifted from the world's poor.

There are opposing views on what should be done. *Study the viewpoints expressed in the letters to newspapers (see the following text box). Then explain your preferred solution to the problem of world debt.*

Third World babies now go hungry because of the need to service national debts. This cannot be right and something must be done about it.

In any case, it should be recognised that the debts of many Southern countries are so great as to be simply unpayable. An unpayable debt is a form of enslavement and we would do well to remember the old Eastern adage that 'one does not owe what one cannot pay'.

I am not suggesting that all debts can be forgiven instantly. But allow me to suggest the following as a possible basis for a way forward:

(a) There should be an end to a net flow of resources from the developing to the developed world. This is needed to promote economic recovery and human empowerment in the South.

(b) Debt repayments should be reduced to a level that allows Third World countries to develop now and in the future.

(c) Developing countries which have their debts reduced should be required to implement and promote the development of their poor and the self-reliance of their nations.

M.B._____Dublin 8

A cancellation of international debts would have one major result – economic ruin and misery in developed and developing countries alike.

The nonpayment of huge Third World debts might very well bring about the collapse of the world's banking system. This would mean not merely the loss of people's hard-earned life savings and pension funds. It would plunge the world into a recession that would prove ruinous to virtually every nation on earth.

Nor would the cancellation of international debts necessarily bring even temporary improvements in the lives of the poor. Corrupt Third World governments would, true to form, take advantage of debt cancellations to buy even more luxury goods for themselves and their cronies. Meanwhile, the condition of the poor would remain as dire as ever.

W.T._____Belfast

Why should the peoples of countries such as Zaire and the Philippines be strapped with debts which they did not incur in the first place?

These debts were offered by foolish bankers to the dictators who held these people in bondage and who were certainly never elected or appointed to office by the people. Now that these dictators are dead and gone, Western bankers and governments would have us believe that the people should be responsible for the debts of their oppressors. The very idea is ridiculous!

Debts are like chains that bind the poor of this world. My motto is this: Break the chains! Scrap the debts! Empower the people!

M.O'C._____County Cork

Activities

1. Describe the origins and effects of the IMF's structural adjustment programmes.
2. What do you understand by the term 'cycle of poverty'?
   Explain how the weight of national debt can give rise to a cycle of poverty in any named Third World country.
3. *'There are varying opinions on what should be done about international debt problems.'* Discuss.
4. Study cartoons A and B left and below.
   (a) What does cartoon A say about international debt?
   (b) How does the message of cartoon B contrast with that of cartoon A?

CHAPTER 8
THE AID DEBATE

Aid is the transfer of money, food, skills or technology from developed nations to developing nations. When properly applied, it can help to empower people by stimulating economic growth and human development in the Third World. Different sources and types of aid are referred to in Figure 8.1.

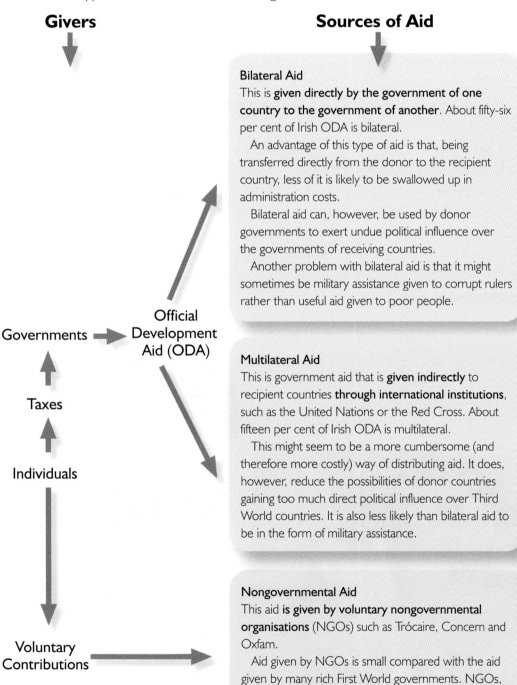

Givers

Governments

↑
Taxes
↑

Individuals

↓

Voluntary Contributions

Sources of Aid

Official Development Aid (ODA)

Bilateral Aid
This is **given directly by the government of one country to the government of another**. About fifty-six per cent of Irish ODA is bilateral.

An advantage of this type of aid is that, being transferred directly from the donor to the recipient country, less of it is likely to be swallowed up in administration costs.

Bilateral aid can, however, be used by donor governments to exert undue political influence over the governments of receiving countries.

Another problem with bilateral aid is that it might sometimes be military assistance given to corrupt rulers rather than useful aid given to poor people.

Multilateral Aid
This is government aid that is **given indirectly** to recipient countries **through international institutions**, such as the United Nations or the Red Cross. About fifteen per cent of Irish ODA is multilateral.

This might seem to be a more cumbersome (and therefore more costly) way of distributing aid. It does, however, reduce the possibilities of donor countries gaining too much direct political influence over Third World countries. It is also less likely than bilateral aid to be in the form of military assistance.

Nongovernmental Aid
This aid **is given by voluntary nongovernmental organisations** (NGOs) such as Trócaire, Concern and Oxfam.

Aid given by NGOs is small compared with the aid given by many rich First World governments. NGOs, however, can operate flexibly and so they sometimes succeed well in helping those most in need. NGOs such as Trócaire and Oxfam also do valuable work in raising awareness within First World communities of Third World problems.

Types of Aid

Development Aid
This is aid spent over a period of time for improving agriculture, health services, educational facilities, etc.

It could come in the form of *financial aid* (grants or loans to be repaid with interest), *personnel* (paid or voluntary nurses, engineers, teachers, etc.) or *technical aid* (tractors, medical equipment, etc.).

The overall purpose of development aid should be to empower Third World people to help themselves.

Emergency Aid
This aid is given in times of crisis to prevent people from dying. It could include food, clothing, temporary shelter or medicine and might be given in response to emergencies such as that caused by the earthquake in El Salvador in January 2001.

Fig. 8.1 Sources and types of aid.

AID – WHO BENEFITS?

International aid is of immense benefit to the less well-off members of our human family. Not all forms of aid, however, are as effective as they might be. Some people argue that the benefits of certain forms of aid are enjoyed by the richer rather than the poorer people of our planet.

Arguments for Aid

Emergency Aid

Emergency aid in the form of food, fresh water, medicines and temporary shelter has **saved countless lives** in situations such as famine in Ethiopia in 1984, floods in Mozambique in 1999 and earthquakes in El Salvador and India in 2001. Modern communications and transport have made the delivery of emergency aid faster and more effective than ever before.

A survey by *Save the Children Fund* investigated the effects of emergency aid during a particularly severe drought in Ethiopia. The survey found that food aid had saved thousands of lives, that it reached even the most remote regions and that it did not damage the livelihoods of local food producers, as many people had predicted it would. The survey concluded that our ability to deliver emergency aid was now so great that, apart from hunger owing to wars, there need never again be a fatal famine in the world.

Development Aid

'*Give me a fish,*' says a Chinese proverb, '*and you feed me for a day. Teach me to fish and you feed me for a lifetime.*'

Development aid usually tries to follow the guidelines of this proverb. It plays an important role in permanently **empowering** people in the South by helping them to help themselves.

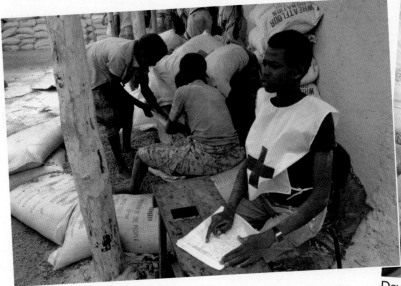

Emergency aid tents.

Development aid.

Contrast the types of aid shown in these two pictures. Describe the importance of each type of aid shown.

Development aid can assist Third World countries to set up vital infrastructure such as water, supplies, sanitation and roads. In that way it can give a 'quick start' to countries where sufficient funds for development are lacking. Most Irish official aid to developing countries is in the form of developmental aid (see 'Ireland Aid' text box).

Ireland Aid

By 2000, Irish governmental aid (known as Ireland Aid) had grown to £178 million. A positive point about Irish aid is that (unlike EU official aid) it is focused on poverty reduction in the world's **least developed countries** (LDCs). Ireland Aid finances programmes in six LDCs in Africa and in East Timor in Asia.

Most Irish government aid programmes are carried out in accordance with local Third World wishes and with the active involvement of local people and governments. The Ireland Aid office in Northern Zambia, for example, is staffed and managed entirely by Zambians.

Irish aid concentrates on developing facilities that will cater for the long-term, basic **needs** of the poorest people. It also focuses on aid that will help to make people **self-sufficient** in catering for these needs. It funds projects such as clean water schemes in Lesotho, farm livestock improvement in Tanzania and primary education and adult literacy courses in Ethiopia. Several of those schemes serve to **empower** local people. For example, health clinics in Zambia train local people in community health care so that they themselves can improve local health standards on an ongoing basis. Aid such as this is referred to as **'appropriate aid'** because it serves the real needs of local people and can be successfully operated by them.

Human Development

Aid can assist human development in First World as well as Third World countries. Aid acts as **an expression of humanitarian concern** and gives opportunities to people in the North to act on that concern. NGOs in particular give people a chance to contribute financially or to become directly involved in assisting fellow members of our human family. Irish NGOs such as *Concern* and *Goal* give suitably trained Irish people opportunities to take part in useful voluntary work in the South. The government's *APSO (Agency for Personal Services Overseas)* programme also provides opportunities for trained, salaried personnel to work on development projects in countries such as Lesotho and Tanzania.

The prejudice shown by some Irish people towards Third World immigrants shows the importance of **development education** as a means of fostering human development at home. Organisations such as *Trócaire*, *Oxfam* and *Afri* provide workshops, seminars and resource packs to inform Irish people on the dangers of xenophobia (hatred of foreigners) and on important, current developmental issues throughout the world.

African LDCs receiving Irish governmental aid are shown in Figure 8.3 on page 66.

Such development work and education greatly assist our own development as caring human beings. They also contribute to world peace and 'interdependence' by forging understanding and **positive bonds** between other peoples and ourselves.

Arguments Against Aid

The Dam and the Hawks – A Story of Tied Aid

In 1996, the then Conservative government of the United Kingdom gave $100 million of tied aid to the government of President Suharto of Indonesia. This aid was used to build a big hydroelectric power station at Samarinda. The power station would provide electricity for those Indonesians who could afford to buy it, but it would also cause large-scale deforestation in one of South East Asia's last remaining rainforests and would result in the local Dayak people being driven from their forest homes.

In return for this 'developmental' aid, President Suharto agreed to purchase twenty-eight Hawk jet fighters which Britain needed to sell in order to finance its arms industry. Suharto was a dictator who suppressed small nations such as East Timor, where his forces murdered up to one-third of all East Timorese. The jet fighters were of potential use to the Indonesian government in its oppression of the Timorese people.

Tied Aid

Tied aid is aid given only on **condition** that the receiver carries out some service for the donor. Typically, a receiving Third World country might be expected to buy goods from an aid-giving First World country. This can be reasonable if the goods are of good quality, fairly priced and of real benefit to the people of the receiving country. Very often, however, this might not be so. In such cases, as the case study illustrates (see 'The Dam and the Hawks'), tied aid can do more harm than good to the country that receives it.

Aid and Politics

The argument has been made that some aid is a legacy of colonialism and is primarily designed not to help the poor, but to maintain the **political influence** of developed countries over developing countries. People point to the following facts to support this belief:

- Most aid is given not to the countries most in need of it, but to relatively **better-off Third World countries** which might serve the political or economic needs of donor countries. A United Nations Development Programme revealed that the richest forty per cent of developing countries got over twice as much aid per person as did the poorest twenty per cent of developing countries. Between 1986 and 1997 the share of EU aid to least developed countries (LDCs) fell from forty per cent to twenty-eight per cent.
- Some powerful countries use aid as a means of **political control** over weaker countries. In 2003, the United States requested the use of Turkish airports as bases from which to bomb Iraq. When the Turkish government refused this request, the US immediately withdrew millions of dollars worth of aid from Turkey.

- During the **Cold War** between communist and capitalist countries, Western aid was used to block the spread of communism. In some cases, it was even used to prevent the survival of democratically elected, socialist governments in the West. When the Chilean people elected a socialist government in the early 1970s, the United States stopped all aid to Chile. The elected Chilean government was later overthrown by an anti-Communist but violently oppressive military dictatorship under General Pinochet. US aid to Chile was then resumed immediately.

When this military coup (takeover of the state) overthrew the elected government of Chile the United States provided aid for the new dictatorship. Why?

- Since the end of the Cold War, aid has been frequently used to encourage **global free trade**, which opens the markets of poor countries to the products of rich nations. This can further enrich the richer nations, but may have devastating effects on the local industries of poor countries. Mozambique, for example, was recently granted aid only on condition that it lowered its import tariffs (taxes) on imported goods. The aid was welcomed in this poor country, but the lowering of tariffs meant that many Mozambiquan businesses were unable to survive and many thousands of Mozambiquans lost their jobs.

False Aid

Some aid can actually do more harm than good to the peoples of the countries that receive it, as the following points illustrate:

- Much international aid is in the form of **loans**. Some of these are soft loans, which can be paid back over long periods and at small rates of interest. Other loans, however, are given on standard commercial terms and contribute to the crippling national debts of many Third World countries. Zambia has to spend four times as much on servicing its international debts than it can on primary education or health care. For every one euro that flows from North to South in the form of aid, nine euro go from South to North, largely in the form of debt repayments. In that sense, aid in the form of loans can contribute to the poverty rather than to the development of Third World countries.

Describe the message of this cartoon.

- Aid in the form of **military assistance** has caused death, destruction and oppression in the South. In the mid-1970s, for instance, the USSR was hoping to gain political influence in east Africa, so it offered military aid to Somalia in its war with neighbouring Ethiopia. For similar reasons, the United States gave military support to Ethiopia. In 1977 the American-backed emperor of Ethiopia was overthrown and replaced by a communist government. The Russians and Americans quickly changed sides, with the Americans now arming Somalia and the Russians assisting Ethiopia. Such 'aid' allowed the war to continue without interruption. Hundreds of thousands of people died in an 'aid'-fuelled conflict between countries that could barely feed their citizens.

Not Enough Aid

Many people believe that one of the main problems with international aid is that there is simply not enough of it. The United Nations recommends that if Majority World poverty is to be combated effectively, developed countries should set aside **0.7 per cent of their annual GNP** for this purpose. Less than one per cent of GNP might seem like a modest price to pay for the reduction of world poverty. Yet **only four countries** – Denmark, Norway, the Netherlands and Sweden – have so far answered the UN's call. Ireland's aid, though growing, reached 0.45 per cent of GNP in 2002. This is well short of the UN target. The United States, which is the world's richest country, has the power to play a critical role in the alleviation of world poverty. By 2000, however, US aid as a percentage of GNP stood at a mere 0.1 per cent.

Fig. 8.2 Aid as a percentage of GNP of selected countries in 2001. Analyse and comment on the information provided in Figure 8.2.

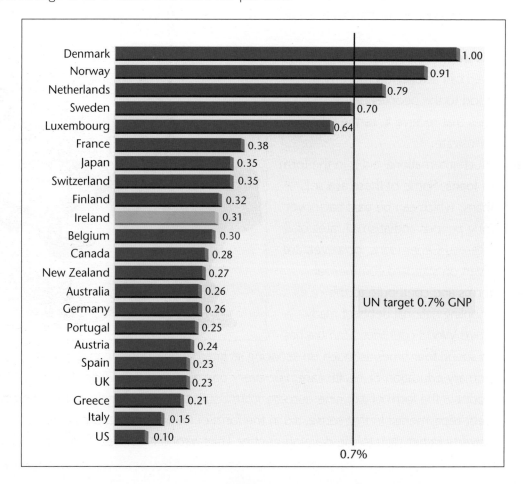

A worrying aspect of EU overseas aid is that at a time when the number of absolute poor people in the world is increasing, **most EU member states are reducing their official aid allocations**. The decline in EU aid to the Third World is partly the result of a belief among European decision-makers that private investment and global free trade can solve the social and economic problems of developing countries. Such a belief seems to ignore the fact that as world trade has increased in value, so has world poverty. It is a somewhat similar belief to that which discouraged the distribution of food aid in Ireland during the Great Famine of 1845–1849 on the grounds that such aid might damage 'free trade'.

THE ROLE OF NGOs

Nongovernmental agencies (NGOs) are private organisations that provide aid to developing countries. These voluntary aid agencies range in size from tiny groups organised locally in developed countries to large multinational organisations. The NGOs most active in Ireland include **Trócaire**, **Concern**, **Goal**, **Afri** and **Oxfam**.

Voluntary agencies can adopt any of four different approaches to aid. An NGO would most likely involve itself in a combination of these approaches while perhaps favouring some more than others:

- **Relief**: All of Ireland's NGOs devote some of their resources to saving lives in emergency situations, such as those caused by the earthquakes in India and El Salvador in 2001, or to providing welfare work without reference to long-term development. Such aid, while effective and vital, delivers relief without confronting the political or social causes of poverty.
- **Development**: Most aid from Irish NGOs is devoted to local, long-term and self-help development projects in Third World countries. Trócaire, for example, supports 538 health, housing, educational and other projects in fifty-six developing countries. Concern has also funded many developmental projects, such as the provision of a low-interest credit scheme that allows people in Bangladesh to develop small businesses without falling into the clutches of ruthless, high-interest moneylenders.
- **Empowerment**: This approach is more political and radical than the two previous approaches. It helps to fight the causes of poverty by empowering poor people to claim their rights, to challenge injustice and to question international inequality. Bodies such as Oxfam and Trócaire in particular have been 'justice driven' in their approaches to aid. Trócaire helped to educate local Mozambican peasants on their land rights so that

An adult literacy class sponsored by an Irish NGO.
In what ways might classes such as these help to empower people?

the peasants were able to prevent a local business person from grabbing their ancestral lands.
- **Education**: Some NGOs recognise that changes in the world economic and political systems of inequality are vital to real and lasting human development. Afri, Oxfam and Trócaire have therefore put considerable efforts into raising public awareness of injustice and of the need for its elimination. Interschool debates organised by Concern have also helped to increase awareness on development issues. In general, though, education and campaigning receive fewer resources than the other approaches among Irish NGOs.

The Strengths and Weaknesses of NGOs

Strengths

Their private nature, **flexibility** and willingness to take political risks can often make NGOs more effective than governments in their approaches to aid. Voluntary bodies can work in areas where, because of local conflicts or international politics, governments could not. NGOs frequently bypass governments and can thus get more directly in touch with the local Third World people and their real needs. This is especially important in situations in which governments do not represent the interests of the poor.

- Because most NGOs are relatively small, they seldom get involved in large-scale, 'inappropriate' development projects such as building huge dams or expensive, specialised hospitals. Instead, they usually involve themselves in locally driven **'appropriate aid'**, which serves the real demands of local communities.
- NGOs can work independently for **political change in the developed world**. They can expose and challenge injustice and inequality, which are the root causes of a great deal of world poverty. A campaign by Trócaire in the 1980s against apartheid in South Africa and a campaign by Irish NGOs in the 1990s to 'break the chains of international debt' are examples of such work for political change.

Weaknesses

- The **scale** of funds in the hands of voluntary agencies is very small compared with the funds that governments can invest in aid. The aid given by Irish NGOs is, on international terms, large in proportion to Irish government aid. Yet in 1988 the combined expenditures of our leading NGOs (Trócaire £26 million and Concern £23 million) came to less than one-third of the value of Ireland's official aid programme.
- In recent years, competition between agencies has led some to place too much emphasis on **self-publicity** to the neglect of the real issues of injustice and Third World poverty. Some voluntary agencies have resorted to the 'helpless starving baby' image as a means of fundraising. This image, though very effective for fundraising in the short term, distorts the First World view of Third World people.

- Some NGOs tend to adopt a **'do gooder' approach** to aid which seems to be driven by the views and emotional needs of people in the North rather than the needs of people in the South. *Child sponsorship schemes* have created controversy even within voluntary bodies themselves. Such schemes encourage individuals in developed countries to provide regular aid for particular Third World children. In return for this aid, donors receive photographs, letters and progress reports from 'their' sponsored children. Many people feel that such schemes do little to promote the dignity or wider social problems of Third World peoples. Individual child sponsorship may also tend to entrench subconscious feelings of superiority among First World donors and of dependence and inferiority among Third World recipients.

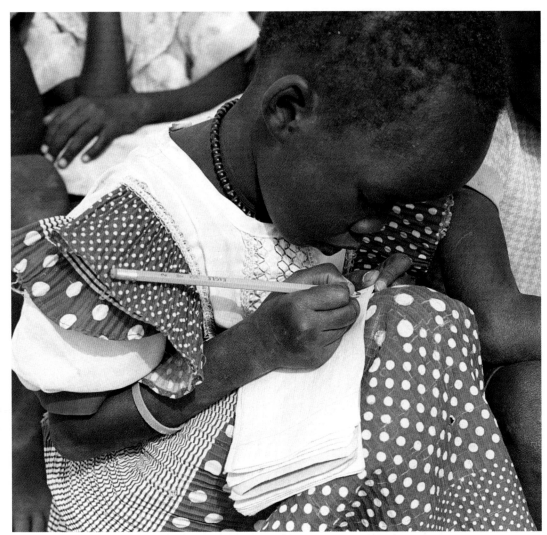

Children who receive individual sponsorship aid are normally encouraged to write regularly to their sponsors. What do you think the strengths and weaknesses of individual child sponsorship are?

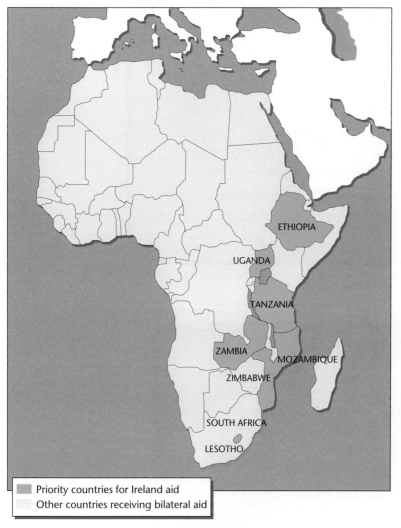

Fig. 8.3 African countries receiving Irish and bilateral aid.

Activities

1. *'World Aid – who really benefits?'* Write an essay on this topic.
2. Describe and evaluate the aid-giving role and effectiveness of Irish nongovernmental organisations.
3. Interpreting data:
   Use the information from each of the pie charts (Figure 8.4) or from the map (Figure 8.3) to write a statistical account of Irish aid.

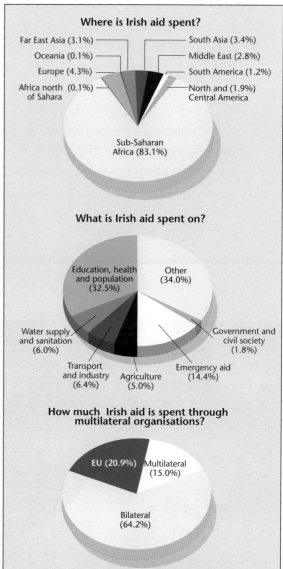

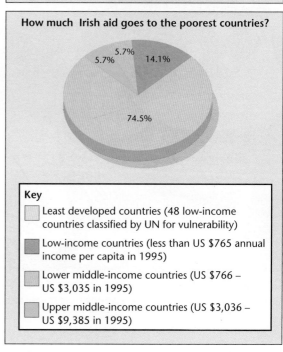

Fig. 8.4 Aspects of Irish governmental aid.

Chapter 9
Empowerment through Land Ownership, Co-operation and Participation

Land ownership patterns often play a key role in rural development. If the people who work the land have a level of control over it, they will have a much better opportunity to develop both economically and humanly.

Case Study: Land Ownership and Empowerment in Nineteenth Century Ireland

During the mid-nineteenth century, land was Ireland's main source of wealth. Most of the land was owned by a relatively small number of wealthy **landlords**. However, a large number of **tenant farmers** and labourers worked the land. This pattern of land ownership created several social and economic difficulties for the tenants who rented the land:

- Some landlords charged rents that were so high that they made the economic lives of tenants difficult. Such **'rack rents'** were especially hard on tenants with small holdings.
- In the mid-nineteenth century, up to one-third of all landlords did not live on their country estates. These **absentee landlords** normally took little interest in farming improvements. In some cases, tenants who improved their holdings were rewarded with demands for higher rent. Such situations worked against economic development and human initiative.
- Tenants usually held their farms by verbal agreement and on a year-to-year basis. They were referred to as **'tenants at will'** because they could be evicted from their homes and holdings at the end of any yearly letting. In 1849 (the final year of the Great Famine) over 13,000 families were evicted.
- Many landlords were fair to their tenants, but tenants were largely **powerless** in the hands of any ruthless landlords. In 1861, landlord John Adare of the Derryveagh Estate in County Donegal threw sixty farmers and their families out of their homes in the depths of winter because he suspected that some of them might have been involved in the murder of one of his agents.

Evicted tenants in nineteenth century Ireland. How did the land ownership system affect social and economic development?

The Land War

In October 1879 Michael Davitt, supported by Charles Stewart Parnell, set up the **National Land League**. The long-term aim of the Land League was to win ownership of tenant farms for the tenants. The struggle that ensued between the Land League and the tenants on one side and the government and the landlords on the other became known as the '**land war**'. It ultimately led to government Land Acts which were to wrest ownership of the land from landlords to tenants.

Tenants at a Land League meeting. Activity in Land League affairs was in itself an empowering process that helped the human development of many tenants.

The Land Acts

Between 1885 and 1905, successive UK governments passed a series of **Land Acts** that essentially transferred ownership of most tenant holdings to tenant farmers. The basic principle of the Land Acts was to set up land **purchase schemes** for the transfer of land. The government would give loans to the tenants to enable them to purchase their holdings (farms) from the landlords. The tenants would then repay the loans through a system of annual payments or 'annuities'.

There were several Land Acts, including the Ashbourne Act (1885), the Balfour Act (1891) and the Wyndham Act (1903). The final Land Act, passed in 1909, made it compulsory for the remaining landlords to sell holdings to tenants who wished to buy. This marked the final phase of a land-ownership revolution in Ireland. Ireland's land was now firmly in the hands of Irish farmers.

Empowerment and Development through Teamwork and Participation – The Co-operative Movement in Ireland

Land ownership gave a degree of empowerment to Irish farmers, but it was not sufficient in itself to lead to significant economic development. Each farmer controlled an isolated economic unit and enjoyed no support network in the formidable task of agricultural improvement. Individual farmers suffered from a lack of capital, technology and agricultural education. The standards of farm produce, such as home-produced butter, varied greatly. The transportation and marketing of produce created great difficulties.

Sir Horace Plunkett set about helping Irish farmers to overcome these difficulties. He realised that farmers needed to work together as a team if they were to improve their situations. By this time, Denmark had displaced Ireland as the main supplier of butter and bacon to the British market. Plunkett saw that the secret of Denmark's success was that Danish farmers had used co-operation as a key to economic development. They had formed agricultural co-operatives that allowed them to revolutionise farming efficiency and to greatly increase the sale of Danish farm produce abroad (see the text box on this page).

Plunkett founded the **Irish Co-operative Movement** on the Danish model. In 1889, he started the first co-operative in Doneraile, County Cork and the movement spread. In 1894, he started the **Irish Agricultural Organisation Society** to co-ordinate the work of local co-operatives.

Plunkett's movement gradually developed so that by 1914 it embraced over one thousand co-operatives. It was most successful in the dairy-farming areas, where small co-operative creameries provided work for local people and improved incomes for their farmer-members. As the twentieth century progressed, the influence of 'co-ops' continued to increase. Co-operatives became fewer in number and larger in size, as small units amalgamated or were taken over by larger co-operatives. These grew into big, successful businesses, which now run numerous processing plants, marts and co-op superstores throughout Ireland, and which export products such as

Co-operatives in Denmark

In the 1880s, Danish farmers began to organise themselves into co-operative societies. These voluntary organisations operated mainly at local level and were **usually owned and democratically controlled by the farmers themselves**. By the 1920s 1,500 dairy co-operatives existed in Denmark.

The co-operative principle spread quickly to all branches of agriculture. Some co-ops set up shops to **buy in bulk** seeds, machinery and other items and then to sell them cheaply to the farmer members. Other co-operatives established **dairies and bacon factories** in which they processed their own produce. These plants accepted only top-quality produce and made sure that finished products only of uniform standard and high quality were released on the market. Danish farm products soon earned such a high reputation for quality that they sold widely and at high prices in countries such as Britain.

As time passed, various types of farmers' co-operatives came to control the entire Danish agricultural industry. **Machinery co-ops** rented or sold the latest farm machinery. Some co-operatives organised **agricultural research**. Others specialised in **marketing** farm products, providing **agricultural education and advice**, **bank loans** or even **insurance** for farmer-members.

Co-operatives came to play a huge role in rural life, with most farmers belonging to several different co-operatives at once. The efficiency of the co-ops played a vital role in the success of Danish farming. Most of all, the co-operative movement **empowered** Danish farmers by helping them to maintain ultimate control over and some level of **active participation** in their own industry.

cheese and butter to many parts of the world. The development of co-operatives has played a vital role in empowering farmers to actively participate in the economic growth of agriculture.

Local Enterprise Boards – Assisting Local Empowerment in Irish Communities

During the 1990s, the government encouraged the establishment of **Local Enterprise Boards** (LEBs) throughout the country as a means of helping local people to develop their own areas, both economically and socially. The LEBs usually take the form of private limited companies managed by representatives of local interest groups such as community councils, chambers of commerce and farmers, as well as by representatives of state bodies such as *FÁS* and *Teagasc*. Many LEBs are active in the following two areas of local development:

- They manage the **Leader Project**, an EU-funded rural development initiative which operates across the European Union. 'Leader' funds up to fifty per cent of capital investment and up to one hundred per cent of training costs for local development projects. Some of these schemes are community-driven and might entail, for example, the refurbishment of community halls. Most projects, however, are privately run and are designed to create private wealth. Critics of the Leader Project might claim, therefore, that the project uses taxpayers' money to enrich those who are already well-off and to further empower those who are already empowered.
- Many LEBs also administer nationally funded **Local Development Programmes**. Such programmes, unlike most Leader projects, are focused on empowering marginalised groups within our society. They include schemes such as those to train disabled people, Travellers or long-term unemployed people for the workplace. They offer practical assistance to such groups in the form of childcare facilities, advice on interview techniques and assistance in the preparation of CVs. Activities such as these assist the human development of marginalised people. They also foster economic growth by preparing much-needed workers for employment.

Activities

1. With reference to Ireland and other European countries, explain how land ownership patterns have impacted economic and human development.
2. 'Agricultural co-operatives have played a big role in the prosperity of farming and in the levels of participation which individual farmers can play in the farming industry.'
   Discuss the above statement with reference to Ireland and one other country.

The Mitchelstown Co-op.
Rationalisation and amalgamations within the co-operative movement have resulted in the dominance within Ireland of a handful of very large co-operatives.
(a) Define 'rationalisation' and 'amalgamation' in relation to the co-operative movement.
(b) Describe how you think rationalisation and amalgamation have affected each of the following:
   Economic growth
   Decision making
   Direct participation by individual farmers within their co-operatives.

This IT course in County Louth was partially funded by the Leader Project.

CHAPTER 10
HUMAN EXPLOITATION

One of the biggest obstacles to human empowerment is the exploitation of vulnerable people by powerful people. This exploitation can happen at global and local levels.

WORLD TRADE AND GLOBAL EXPLOITATION

All countries must trade in order to prosper and world trade, which is now worth the equivalent of more than €35 trillion each year, is one of the strongest links between countries of the South and the North. But global trade has tended to favour the interests of the wealthy and powerful, rather than those of the poor and vulnerable. The richest twenty per cent of the world's people control eighty-four per cent of global trade, while the poorest twenty per cent of people control less than one per cent of trade. Multinational corporations such as Volkswagen have annual turnovers of about twice the GNP of Bangladesh, while the turnover of the Nestlé Corporation is almost twenty times the GNP of Nicaragua.

Some people argue that Third World poverty has increased because of, rather than in spite of, growing trade. They point to the following problems experienced by many Third World countries.

1. **Reliance on Unprocessed Commodities**
   Ever since colonial times, countries of the South have been used to providing commodities such as coffee, tea, sugar and cotton for the manufacturing nations of the North. These products are usually in their cheapest, raw state and up to ninety per cent of the profit comes from the processing, marketing and retailing of the products, activities which have usually been carried out in First World countries. Prices are usually controlled by powerful transnational corporations (TNCs), such as Chiquita (bananas) and Nestlé (coffee), whose headquarters are all in rich First World countries. Some Third World regions are so reliant on a single commodity that a fall in the price of that commodity can have disastrous economic results. The island of Negros in the Philippines has been so dependent on exporting sugar to the United States that it suffered widespread famine when the US suddenly slashed its sugar imports in 1985.

2. **Unfair Terms of Trade**
   The prices of developed countries' exports to the South have risen enormously owing to inflation, but Third World countries have not received corresponding price

rises for their products. Thus, desperate to sell their goods at almost any price, they have sometimes found themselves having to produce more and more in order to pay for the same amount of manufactured goods from the North. In 1972, for example, Uganda had to sell six tons of cotton to purchase one European truck. By 2002, more than thirty-five tons of cotton were needed to buy a similar truck. These terms of trade (as the relationship between the price of goods being bought and sold is called) are unfair and contribute to the economic exploitation of Third World peoples.

3. Fluctuating Prices

The prices of most Third World commodities, as well as generally being low, have been allowed to fluctuate wildly according to supply and demand on the world market. This has made it almost impossible for Third World countries to plan their economies. The price of sugar, for example, actually fell below the cost of production in March 1992. This resulted in hunger and hardship in the Philippine island of Negros. The World Trade Organisation (WTO) has sometimes contributed to fluctuating prices by encouraging the overproduction of commodities in the South. Between 1980 and 1992, the WTO encouraged some west African countries to increase cocoa exports in order to pay international debts. The resulting increase in exports resulted in the collapse of world prices. Consequently, countries such as Ghana doubled their exports, but actually earned less foreign exchange (see Figure 10.1).

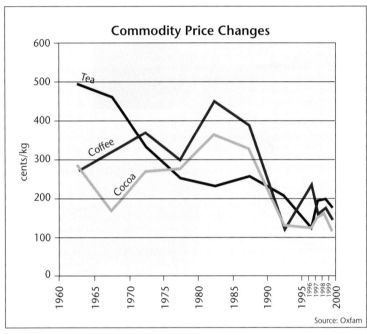

Fig. 10.1 Commodity price changes.
Describe and comment on the likely socio-economic effects of the price trends shown here.

Fact File on Coffee – A Leading Third World Commodity

Coffee is the world's second-most important commodity after oil. It is produced mainly in Central and South America and in Africa. Some countries, such as Burundi and Ethiopia, depend on coffee for most of their export earnings. A bad harvest or a sudden drop in coffee prices can bring bankruptcy to such countries.

The price of coffee was once regulated on the world market through an International Coffee Agreement (ICA), but the insistence on global free trade by rich coffee-consuming countries and by the World Trade Organisation brought an end to price control. Wildly fluctuating prices have resulted. Once in 1989, the price of coffee fell by one-third in a single week!

Only a tiny proportion of the price one pays for coffee goes to the Third World workers that produce it. Much more goes to advertising and profits for the Western-owned TNCs that control the coffee trade.

4. Cheap Manufacturing Labour

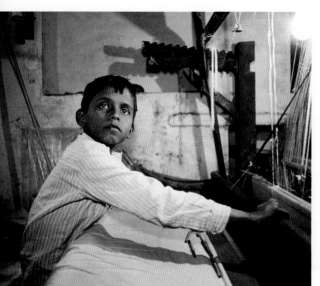

Workers such as this Indian child provide virtual slave labour for wealthy textile manufacturers.

The tradition of the South only producing primary products and of manufactured goods being processed only in the North has changed considerably in recent times. Many manufactured products, such as clothing and footwear, are now being produced increasingly in China, Indonesia, the Philippines and other Third World countries. The emergence of new technology and TNCs has been largely responsible for this trend. New technology has reduced the need for highly paid skilled labour. It has allowed TNCs to abandon traditional and relatively expensive First World manufacturing locations in search of less skilled but cheap labour in the South. TNCs bring much-needed employment to some Third World communities, but they have sometimes reaped huge profits on the backs of poorly paid and badly treated employees (see the text box below). Most of the profits return to the First World countries where TNCs have their headquarters.

Sample Study – Sweatshop Labour in South East Asia

Next time you slip on a pair of brand-name running shoes, give a thought to who may have manufactured them and where. Some have been produced in Indonesia by the workers of Jabotabek, who have made runners, Y-fronts and jeans for some of the West's most famous brand names. Jabotabek is Indonesia's showcase industrial complex, where Taiwanese and South Korean companies use cheap labour to manufacture shoes, shirts and textiles under licence to Western label-owners who buy back the semi-finished products. They then mark up the prices and sell them to the wealthy European and American markets ... 'A $130 pair of sandshoes cost only a few cents,' says a British industrial relations specialist who recently toured Jabotabek.

In Indonesia the legal minimum wage is $1.27 a day, but 12,000 factories pay less than forty per cent of the minimum wage. Female workers comprise eighty per cent of the labour force. In many factories wages are cut if employees go to the toilet unauthorised, and pregnancy means instant dismissal. A normal working week is more than fifty hours, with no payment for overtime. Workers who complain or organise a strike are often interrogated and electrocuted by the military. The names of people who are sacked are entered on a blacklist which is circulated to employers, removing any chance of getting another job.

From *Pathways to Geography*, Paine and Bliss, Melbourne

Activities

1. (a) What is meant by the term 'sweatshop'?
   (b) In your opinion, why do the employers referred to favour the use of female workers?
   (c) List the ways in which the employees referred to are being exploited.
   (d) In your opinion, why is human exploitation such as that described above allowed to continue at present?

EXPLOITATION AT LOCAL LEVEL

Case Study 1: Sex Slaves Reach Ireland

It is estimated that criminal gangs have illegally trafficked more than 100,000 women into Europe, and a small minority of these women into Ireland. These women come from Eastern Europe, the Middle East and Africa. They start off as economic migrants who hope to secure better lives in Europe, not originally understanding that this is legally impossible without proper visas and work permits.

Members of the international Mafia and other **criminal gangs** promise and even sign 'contracts' guaranteeing jobs in European restaurants, bars and offices to unsuspecting Third World and Eastern European women. They smuggle the women into Europe who then live in circumstances of extreme exploitation, as described in the following text boxes.

Gang members confiscate their victims' passports and hold the women under threats of violence. Lacking legal documents, local knowledge and language skills, the frightened women find it almost impossible to resist or escape from their 'minders'. They essentially become **slaves**.

About one-third of the slave-women become locked into **prostitution**, particularly in the Netherlands and Belgium. Approximately another third are trafficked into '**domestic service**' in which they work for a pittance for rich European families. Most of the rest end up working in **illegal sweatshops**, particularly in the textile industry in Italy.

Development agencies such as Trócaire and the Rahuma Women's Project estimate that more than 100 female sex slaves have ended up **in Ireland**. They are believed to be virtually imprisoned in 'closed' houses that function as brothels. Some are believed to have become mentally deranged because of the coercion and exploitation that they suffer. A few have escaped or have been rescued by bodies such as the Rahuma Women's Project. They live in fear of being deported to their home countries, where they could be in danger from the Mafia gangs who originally trafficked them.

A prostitute in Amsterdam, the Netherlands.
A minority of female prostitutes in Europe have been virtually enslaved by criminal gangs that traffic them into Western Europe from Third World or Eastern European countries.

Case Study 2: The Boys at Letterfrack

An imposing and remote building in County Galway once housed up to 200 boys at any one time. They were the inmates of Letterfrack Industrial School, which had been set up in 1887 to discipline and mould unruly boys, generally from poor families. The boys were aged from six to sixteen. At the reform school they were supposed to receive the education and training needed to make them productive and responsible members of society.

Local people remember Letterfrack as a self-sufficient industrial entity with its own farm, bakery, carpenter, blacksmith and tailor. The letterhead on the school's stationery proudly announced 'Orders Received in Tailoring, Bootmaking, Carpentry, Bakery, Cartmaking, Smithwork' and other services.

Behind this image of good order and moral reform lurked another side to the industrial school. Boys as young as six were made to work long hours and for no pay at the various 'industries' which made up to £12,000 annual profit for the school. Formal education took second place to forced labour. Discipline was rigorous and beatings were common and allegedly severe. Between 1996–2001, more than 150 allegations of physical and sexual abuse were made against the institution by former inmates who lived at the school until its closure in 1974.

Even allowing for the bygone educational belief that 'to spare the rod is to spoil the child', it would appear that institutions such as Letterfrack wittingly or unwittingly took part in severe human exploitation at local level.

Analyse the view of this cartoon on the aspects of international trade.

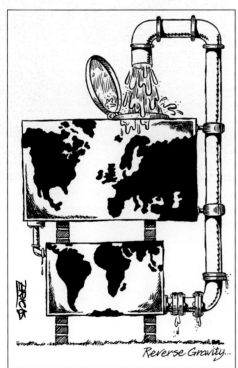

Reverse Gravity...

Case Study 3: Bonded Labour in Pakistan
For a study of bonded labour, see pages 100–103.

Activities

1. State briefly but clearly the meanings of each of the following terms:
   - TNC
   - Terms of International Trade
   - The International Coffee Agreement
   - Sweatshop labour.
2. Discuss the view that international trade as it now exists is a source of human exploitation at a global level. Refer in your answer to the effects of trade on the peoples of specific countries or regions.

   Or

   Evaluate the validity of the message in the cartoon on this page with reference to the workings of international trade.
3. Human exploitation at local level can prevent the empowerment of many individuals. Describe the workings and effects of two or more examples of such exploitation.

CHAPTER 11
GENDER ROLES

Discrimination

Poverty, injustice and tradition have often conspired to place females in subservient positions to males. This discrimination against women has seriously hindered development, particularly in many Third World societies. It has taken many forms in different parts of the world:

- Some women suffer **discrimination under the law**. In certain Moslem societies, daughters have been allowed to inherit only half as much as sons. In others, women have been forbidden to initiate divorce against husbands, while men can quite easily divorce their wives. Throughout the 1990s, the Taliban government in Afghanistan compelled women to heavily veil themselves when appearing in public. It also forbade women to attend schools or work outside the home, where they might mix with men. 'If women want to study,' declared one Taliban follower, 'they can do so at home under their husband's supervision.'

- In some parts of the world, females are discriminated against at birth or even before it. In China, the government's encouragement of 'one child per family', together with an old Chinese preference for boys who continue 'the family name', has led to the **selective abortion** of many unborn baby girls. In the Madurai district of southern India, some parents feel that

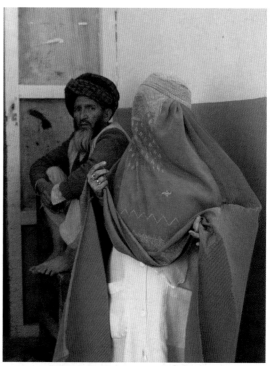

A woman in Afghanistan under the Taliban government.
Discuss why rulers might want women to dress like this.

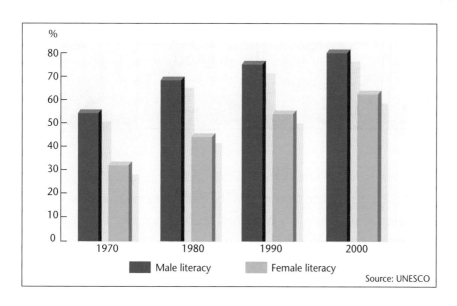

Fig. 11.1 The literacy gap between males and females in the Third World, 1970–2000.
Describe the gaps and trends shown.

> Find out more about discrimination: an issue of gender in *Our Dynamic World 2* Chapter 7.

they cannot afford to bring up daughters and provide them with marriage dowries and other traditional support. In such cases, desperately poor parents have been known to **kill their baby daughters** within hours of birth.

- Throughout childhood, discrimination against females often continues. Where scarce **educational facilities** present economic challenges for poor people, boys tend to be given priority over girls. Figure 11.1 shows that although the global literacy gap between the genders is closing, males are still more likely than females to be able to read and write.

- While Third World boys are given at least the rudiments of an education, girls are often prepared for marriage and motherhood. In some areas of the Ivory Coast, girls as young as eleven years of age are still forced into **arranged marriages** with distant relatives who may be two or three times their ages. Such marriages are arranged by parents to strengthen clan relationships and to 'protect' girls from the romantic 'adventures' which they might otherwise encounter in single life.

- As wives and mothers, women are often burdened with a traditional **triple workload** of caring for children, managing the household and working at tasks such as subsistence farming. While such women work very hard indeed, their efforts are usually **unpaid and even unrecognised** as 'real work' (see the cartoon below).

'The lie of the land.'

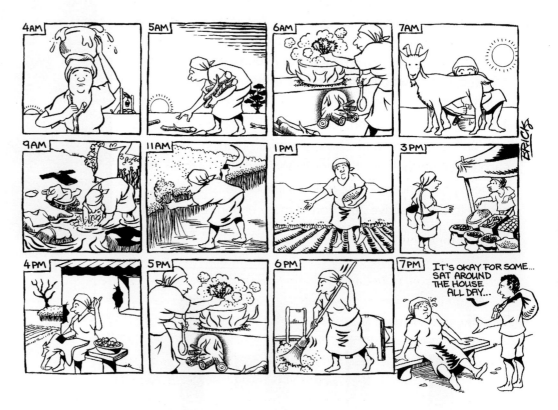

- Many women take part in 'formal work' of the kind that is recognised as contributing to the gross national product (GNP) of a country, but these women are often discriminated against. In many countries, women are paid less than men and are used by transnational corporations as a source of particularly **cheap labour** (see the text box on the next page).

Even in EU member states, women workers are sometimes particularly exploited. In Britain alone there are over one million females who carry out industrial tasks within their own homes. These 'homeworkers' sew garments or pack products for payments as low as £1.50 per hour. They work only when their employers need them, have no trade union to protect their interests and enjoy none of the minimum legal protection afforded to factory workers.

Barbaric Conditions

Female workers in parts of Asia must endure barbaric conditions for a meagre $5 a day. And the West, Abigail Haworth reports, is reaping the benefits of their slave labour.

Marsinah was murdered because she asked for too much. The 25-year-old Indonesian factory worker was dissatisfied with the pay she received for working a 12-hour day hunched over an assembly line performing tedious, spine-damaging work. So she led her fellow workers out on strike to demand a better deal: an increase in daily wages from the equivalent of $1.20 to $1.50, and a 30-cent lunch allowance. Marsinah didn't think her request was too unreasonable.

She was wrong. The day after the strike, which took place on Java, Marsinah disappeared. Her barely recognisable corpse was later found in a shallow, mud-swamped grave . . .

Marsinah's death may be an extreme example, but it is just one of the many dirty secrets behind South East Asia's dazzling economic success. While Western governments applaud the region's dynamic growth rates, little is heard about the hundreds of thousands of young women whose thankless toil makes it all possible.

Herded into the vast Export Processing Zones (EPZs), which have sprung up everywhere from Indonesia to Taiwan, these women work in appalling and dangerous conditions, churning out clothing, electronic products and household goods that they can only dream of owning themselves.

EPZs are increasingly the backbone of Asia's export-orientated economies. Free of tax and red tape, the sprawling industrial areas act as giant magnets to foreign investors fleeing from the profit squeeze in their own countries. Their most alluring feature is cheap labour. Between eighty and ninety per cent of workers in EPZs are young women who, as one Department of Trade pamphlet in the Philippines trumpeted, possess the prized qualities of 'natural subservience' and 'a high tolerance for boredom'. Women who don't fit this mould are dealt with severely. Instant dismissal and harassment are constant threats; imprisonment, torture, and death are all too frequent occurrences.

(Time Magazine, October 1995)

(a) What are the Export Processing Zones?
(b) Outline the injustices women suffer within Export Processing Zones.

WAYS OF CHANGING GENDER ROLES

Self-Help

Self-help is a vital key in the struggle to achieve equal rights for women. A perception has existed that females are generally less militant and more subservient than males, which has contributed to the exploitation of women by ruthless employers and conservative rulers. When women began to organise themselves to assert their social rights and improve their economic situation, this perception and exploitation began to decline. The best self-help initiatives have usually been carried out **at local levels by women working in co-operation** with each other. An example has been the Kassassi Women's Agricultural Development Association (see the text box below).

Kassassi – A Case Study of Women's Self-Help

Kassassi is an isolated village in the African country of Sierra Leone. It is a poor area. The main economy is subsistence farming, and the main 'export' has been young people who migrate to far-away towns and cities.

In an effort to create a better future for their children and to improve their own economic situation, a group of women got together to drain and clear a nearby swampland. Calling themselves the **'Kassassi Women's Agricultural Development Association'**, they joined forces to reclaim and farm about 100 acres of land. The harvests of this land are owned communally and sold to members at a fair price during the 'lean' season when no crops grow.

Although the project is run by the women, it benefits the entire local community. Boys and girls, for example, earn money to pay for school fees by working on the Association's land. For the women involved, the benefits have not only been economic. They have grown in self-confidence and now take part much more in decision-making within the village. Men, too, have benefited from the growing realisation of the organisational and economic capabilities of their women folk. Progress for the women's co-operative has been progress for all in Kassassi.

Education

As mentioned already, parents in many Third World countries tend to spend their limited resources on educating their sons. Girls may be kept at home to help their mothers, who themselves have already been denied education. These situations create cycles of female illiteracy, which in turn reduce female self-esteem and socio-economic expectations.

- **Improved educational facilities** for females are needed to break such cycles of illiteracy and passivity. Many developing countries have made big efforts to provide education for girls as well as for boys. Vietnam is one of the world's poorest countries, yet over eighty per cent of its females are literate.

- Some countries such as Kenya have offered scholarships and reduced school fees to females, but the education which girls receive often lacks the **vocational elements** required to enter skilled employment. As in developed countries, up to ten times more boys than girls received apprenticeship-type training in subjects such as carpentry, plumbing or mechanical engineering. This severely limits women's capacity to enter the workplace on an equal footing with men.

- **Adult education** is very important to the human and economic development of women who were denied the benefits of an adequate formal education as girls. Mothers who learn to read and write or to master other educational skills often discover new levels of awareness and empowerment. They in turn encourage education for their daughters, thus breaking the cycles of illiteracy and impotency of earlier generations.

- Social education aimed at **men** is another vital element in the struggle for gender equality. In developed as well as developing countries, males still control most positions of political and financial influence. For example, men control ninety-seven per cent of the senior positions in the United Nations and more than ninety-seven per cent of the world's property. It is important that men, being in such positions of power, should all come to appreciate the enormous economic waste (and social hurt) that must result from denying one-half of the human family the opportunity to reach its full potential. Such a realisation might encourage many powerful men to abandon conservative ideas on the supposed roles of women and instead to use their influence to support the cause of full gender equality.

Empowerment through appropriate adult education. These Japanese women are farmers who are being trained in agricultural pest management. Once trained, they will be expected to pass on this knowledge to other women and men in their villages.

PLANTU
LE MONDE
Paris
FRANCE

When do we eat?
(a) What are the messages of this cartoon regarding gender issues and education?
(b) Is the cartoon itself fair and appropriate? Explain your viewpoint.

Workplace Reforms

As mentioned already, an increasing number of women are severely exploited at work outside the home:

- The creation of strong **trade unions** for all workers has in the past protected many people in Europe from the worst elements of economic exploitation. The spread of effective trade union activity would greatly help to empower female as well as male workers throughout the world. Such activity, however, is not always easy to achieve. Some Third World governments work hand-in-hand with powerful employers to discourage effective trade unionism. China, for example, permits workers to join only 'official' trade unions that demand little from multinational employers. Many transnational companies avoid having to negotiate with any trade unions at all by subcontracting their work to generally non-unionised subcontractors. The subcontractors who can produce at the lowest costs (usually by paying the lowest wages) tend to win the contracts. The American sportswear company Nike, for example, has manufactured some of its products through subcontractors in Asia.

Nurses demanding improved hospital conditions in Ireland. Discuss the importance of and problems facing trade unions in today's 'global economy'.

- Since the Employment Equality Act of 1997, equal pay for equal work has become the norm in Ireland, yet there still exists an average gender pay gap of fifteen per cent. This occurs partly because men are more likely than women to seek promotion within the workplace. Women might feel better empowered to seek work promotions if **maternity leave and crèche facilities** were more generously and evenly available throughout the country. Lingering **cultural conceptions** of what constitutes 'male' and 'female' work roles also help to account for the gender pay gap. Some Irish men and women have yet to be fully liberated from notions such as those that tend to equate women with housework and men with jobs in carpentry or plumbing. Formal and informal education, ranging from classroom lessons to television advertisements, could do much to achieve this cultural liberation.

What is the cartoon's message?

Gender-Appropriate Aid

In the past, even international aid tended to discriminate against women in developing countries. The traditional European notion of the male 'breadwinner' and female 'housewife' affected many development projects in the 1960s and 1970s. Many aid agencies assumed that if the head of a family (who was almost always assumed to be male) received aid, its beneficial effects would 'trickle down' to the women of the receiving household. Most farmwork in countries such as Kenya and Zambia is carried out by women, yet agricultural assistance to these countries, in the form of educational programmes and technology, almost always went to men. Such aid merely cemented already existing tendencies towards male domination, since education in all its forms tends to empower the receiver.

During the 1980s, **development aid became more associated with gender issues.** By 1995, the world's principal aid donor – the EU – passed a resolution that 'women should participate in and benefit from the development process and that the economic empowerment of women ... must be strengthened.' By the year 2000, development aid had become better designed to facilitate the economic empowerment of women. The Grameen Bank in Bangladesh is an example of such aid. This agency gives loans to more than one million borrowers, most of whom are poor women. Borrowers need no collateral (guarantees) against these loans and the bank's officials are required to live among the people to create a bond between them and the borrowers. Many women have used their loans to buy seeds, oxen, farm machines and eventually land. They have thus broken the cycles of poverty and dependence in which they once lived.

Activities

1. (a) Analyse and contrast the work roles of men and women shown in Figure 11.2.
   (b) Discuss what the information shown on the graph suggests about the varying gender roles in Africa.
2. 'Despite the ongoing improvements in the status of women, social and economic discrimination against females continues to exist.' Discuss.
3. With reference to actual examples, identify three ways in which women have been discriminated against with regard to their roles in society. In the case of each way identified, describe how such discrimination can be brought to an end.
4. Describe specific ways in which the empowerment of women could diminish existing different gender roles in society.

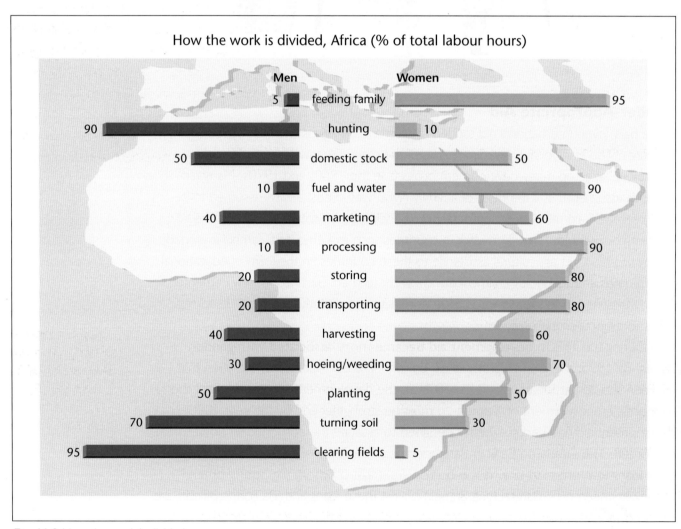

Fig. 11.2 How the work is divided.

CHAPTER 12
THE SUSTAINABLE USE OF RESOURCES

Economic development has often been achieved at great cost to our global and local environments. As described in Chapter 5, it has contributed to deforestation, desertification and the threat of global warming. A **preferred model for future economic and human growth must lie in the sustainable use of economic resources.**

> **Definition**
> **The sustainable use of economic resources** means meeting the needs of the present without compromising the needs of the future. It demands that we utilise the earth only insofar as we respect the rights of all its inhabitants and do not squander resources that will be needed by future generations. Fossil fuels, forests and fish are three examples of resources that need to be used in a sustainable manner.

THE SUSTAINABLE USE OF FOSSIL FUELS

We have seen in Chapter 5 that **global warming** is a major environmental threat to our planet. The overuse of fossil fuels, especially in economically developed countries, is a cause of global warming. It therefore follows that a reduction in the use of fossil fuels is of major importance to sustainable development.

The Kyoto Protocol

Throughout the 1990s, more and more countries saw the need to combat the dangers of global warming. At the UN climate conference in December 1997, held in Kyoto, Japan, the industrialised world agreed an historic protocol (agreement) to reduce greenhouse gas emissions in order to protect the environment. **The Kyoto Protocol** set a target of an overall five per cent reduction in the production of greenhouse gases between the years of 1990 and 2012. This target represents the first tangible international effort to combat climate change.

By the year 2000, however, cracks began to appear in the international resolve to reduce greenhouse production. By that time, it was clear than many countries (including Ireland) had already exceeded their 2012 commitment target (see Figure 12.1). Then, at a meeting at **The Hague** in the Netherlands, it became clear that leading polluting countries wished to water down their Kyoto commitments. The United States and Japan demanded to be given the right to purchase 'pollution quotas' from less developed and therefore less polluting countries. Such a 'trade' would allow rich countries to buy from the poor the right to carry on polluting more or less as normal.

> Find out more about Kyoto in *Our Dynamic World 2* Chapter 22.

In March 2001, the United States shocked the rest of the world by announcing that it was abandoning its Kyoto commitments because a reduction in its greenhouse gas emissions would 'harm the American economy'. It is generally recognised that without the co-operation of the world's leading polluting country, the Kyoto Protocol cannot survive effectively.

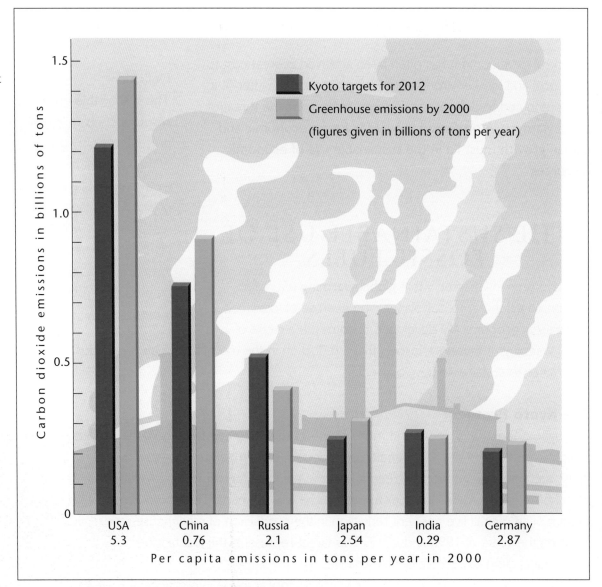

Fig. 12.1 Carbon dioxide emissions in the principal carbon dioxide producing countries, set against Kyoto targets

Ireland and the Kyoto Protocol

Ireland agreed at Kyoto to keep its 2012 emissions to just thirteen per cent above its 1990 level. However, the Irish economic boom had already pushed Ireland's emissions to twenty per cent above the 1990 figure by the year 2000. It seems highly unlikely, therefore, that Ireland will keep its Kyoto commitment. Nevertheless, the government has set out the following **plan of action** in an effort to keep our fossil fuel usage and our greenhouse gas emissions to sustainable levels:

- It is planned to phase out or convert to natural gas the big coal-burning ESB power station at **Moneypoint** on the Shannon Estuary. It is also intended that in the future, more of our energy will come from 'clean' and renewable sources such as wind, and a large wind farm is being planned on the Moneypoint site.

- Efforts will be made to reduce the huge amount of carbon dioxide being produced by our ever-increasing **road traffic**. The *Vehicle Registration Tax* favours vehicles that use less fuel, while the *National Car Test* weeds out old fuel-hungry engines. There will be more investment in *public transport* in an effort to reduce the number of car journeys being taken. *Taxes on fuel* may also be raised to discourage the overuse of private vehicles.

- Our **homes** will also have to change. *New building regulations* are reducing energy consumption in new houses. The government plans to introduce *'Energy Efficiency Certificates'* for houses that were built before 1991. People wishing to sell a pre-1991 house will have to obtain such a certificate in order to do so.

- **Trees** help to reduce the amounts of carbon dioxide in the atmosphere. In the early 1990s, the Department of Forestry pursued a policy of increasing *new forest planting* in Ireland by fifty per cent. *Grants* and other encouragements are now being offered to farmers to grow more trees on their farms. Another greenhouse gas – methane – is produced by cattle as they digest grass. A ten per cent reduction of Ireland's **herd size** is being planned. This will help to reduce greenhouse emissions in agriculture, which at present produces more greenhouse gases than any other economic activity in the state.

Ireland will be fined 1m per year if they don't get their emissions sorted out.

The ESB power station at Moneypoint.
(a) Why is this particular power station being singled out for closure or modification in the fight against global warming?
(b) Suggest some reasons why you think this location was chosen for this power station.

THE SUSTAINABLE USE OF FORESTS

Find out more about Norrland in Our Dynamic World 1 *pages 239–41 and deforestation in* Our Dynamic World 2 *Chapter 25.*

We have learned in Chapter 5 that deforestation is a contributing factor in the creation of global warming and desertification. It also results in the extinction of countless species of plants and animals and threatens the existence of forest peoples such as the Yanomami of Brazil. It follows that the sustainable use of forests is important to the welfare of our planet.

The sustainable use of forests might entail the **conservation** of some forest areas in their present state, the controlled or '**managed**' **use** of other forest areas and the **afforestation** or replanting of trees in areas that have been deforested already, but not yet eroded or desertified beyond recovery.

Case Study 1: Managed Forests in Scandinavia

Forestry is an important source of employment and raw materials in the Scandinavian countries of Sweden, Norway and Finland. These countries have long since moved from a policy of mere exploitation of forests to one of **sustained yield**. In Sweden, over 200,000 hectares of forest are cleared annually. Of this area, over seventy per cent is *replanted* with nearly half a million tree seedlings each year, while natural regeneration is allowed in the remaining area. Careful *disease control*, improved land *drainage* and the application of *fertilisers* also play an important role in the sustained development of Sweden's forests. These forests are managed so that growth exceeds cutting, allowing Sweden's forest stocks to actually increase in recent years.

Case Study 2: The Korup Project of Cameroon

The government of Cameroon in west Africa embarked on a sustainable development plan for the Korup region, which might serve as a model of **sustainable forest use** for parts of the Amazonian or other rainforest regions. The Korup project is an **integrated** one, combining forest conservation with sustainable economic development for local farmers.

About 125,000 hectares of Korup have been set aside as Cameroon's first **national park**. This area, which contains thousands of plant species and many endangered animals, is completely protected

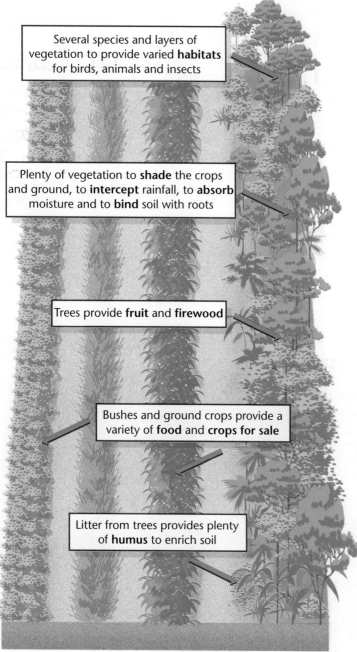

Fig. 12.2 Some advantages of agriforestry.
Describe the advantages of agriforestry from the point of view of the ecosystems, soil fertility and the needs of local people.

Labels:
- Several species and layers of vegetation to provide varied **habitats** for birds, animals and insects
- Plenty of vegetation to **shade** the crops and ground, to **intercept** rainfall, to **absorb** moisture and to **bind** soil with roots
- Trees provide **fruit** and **firewood**
- Bushes and ground crops provide a variety of **food** and **crops for sale**
- Litter from trees provides plenty of **humus** to enrich soil

from deforestation. The use of the national park for controlled and limited tourism could attract valuable foreign revenue to Cameroon.

The immediate social and economic needs of local people have not been overlooked. A **buffer zone** has been created around the protected area. Every effort has been made to sustainably improve productivity within the buffer zone, so that 30,000 local small farmers can make a reasonable living without further forest clearances.

Agriforestry, which involves rotated food crops with trees, has been encouraged among the farmers so that the trees continue to protect the soil and crops from heavy rains. Commercial use has been made of **local forest products**, such as gum, resins and fruits. These products, which are exploited on a renewable basis, provide useful supplementary work and income for local people.

THE SUSTAINABLE USE OF FISH STOCKS

Improved technology and larger fishing vessels have led to serious **overfishing**. This overfishing, carried out mainly by the fishing fleets of the EU, Japan and the United States, has led to a serious **depletion** of fish stocks in many parts of the world. By 2000, for example, two-thirds of the fish stocks in EU waters were approaching commercial extinction.

Traditionally, the governments of developed coastal nations have encouraged their fishing fleets to expand in size and 'efficiency'. It is now widely recognised that such unbridled expansion may destroy the fish stocks on which fishing depends. The priority must now be to ensure that fishing is carefully managed so that it allows fish stocks everywhere to replenish themselves. Fishing, like other commercial activities, must be carried out in a **sustainable** manner.

> Find out more about fishing in *Our Dynamic World 1* pages 210–12 and in *Our Dynamic World 2* Chapter 23.

Case Study: Measures Used to Protect Celtic Sea Herring

By the 1970s, overfishing by Irish and foreign trawlers had almost wiped out herring stocks in the Celtic Sea. Between 1977 and 1982, all herring fishing had to be suspended to allow stocks to recover. The following conservation measures are now used in an attempt to manage stocks and so prevent future stock depletion:

- The European Union sets upper limits or **quotas** on the quantities of fish that each of its member states may catch. These quotas vary according to the stock size of any given fish species in any given year. Each year, a *total allowable catch* (TAC) is agreed for each species. Each member state is allocated a *national quota* or percentage of the TAC. An *individual fishing vessel* is then licensed to catch a small part of their country's national quota.

  Disagreements often emerge on the size of TACs or of national quotas because coastal nations try to promote the interests of their own national fishing industries. The Irish government, for example, complains that while seventeen per cent of all EU fishing waters are off Ireland, the Irish national quota amounts to only 4.4 per cent of the overall TAC.

A Japanese trawler arrested by the Irish Navy for illegal fishing in Irish waters. Poaching presents a major problem for the enforcement of fishery exclusion regulations. With only eight fishery protection vessels to police our fishing grounds, large boats from Spain, Japan and other countries sometimes fish illegally and thereby damage fish stocks within Irish waters.

- **Fishery exclusion zones** have been set up to limit international fishing on shallow, rich fishing grounds near coasts. Waters within ten kilometres of the Irish coast may be fished by Irish boats only. Waters between ten and twenty kilometres of our coasts may be fished by Irish and other EU vessels only. The boats of non-EU countries must limit their activities to outside the twenty-kilometre zone.
- The EU proposes to reduce the size of its **fishing fleet** by forty per cent to lessen the pressure on fish stocks. Despite the seemingly obvious merits of this proposal, it has drawn *criticism* from the fishing industries and governments of individual member states. The Irish government, for example, complained that a reduction in the capacity of Ireland's fishing fleet would be unfair because it possessed less than three per cent of Europe's total 'catching power'.

What About Third World Fish Stocks?

Declining stocks and more rigorously policed coastal exclusion zones have now greatly reduced fishing opportunities in temperate waters. This has caused the world's larger fishing nations to look to Third World fish supplies to maintain unsustainable fishing practices. The EU has made **'Third Country Fishing Agreements'** whereby it can purchase from the governments of developing countries the right to fish in their coastal waters. Greenpeace and the Worldwide Fund for Nature describe these agreements as little more than bribes to poorer countries to allow the pillaging of the world's fish stocks. Some developing countries agree. The governments of Morocco and Chile have each complained that Third Country Fishing Agreements, to which they have been party, have resulted in serious overfishing within their territorial waters.

Are large factory ships such as the *Atlantic Dawn* an advantage to or a problem for the world's fishing industries?

The *Atlantic Dawn* – Europe's largest fishing vessel – was enthusiastically launched by Ireland's Minister for the Marine in February 2000. This 145-metre-long Irish ship cost a staggering €63 million to build. It was financed by a syndicate of Irish banks. The ship was destined to fish, not in Irish or European waters, but off the shores of Mauritania in west Africa. The *Atlantic Dawn* can catch more than a ton of fish every five minutes. Its catch is processed and frozen on board before being sold.

A Senegalese fisherman. Over half a million people in the west African country of Senegal depend on fishing for their livelihoods. Some of these people complain that local fish stocks are being depleted and their livelihoods are being threatened by the large European boats. Under the terms of an EU Third Country Fishing Agreement, European vessels are permitted to fish off the coast of Senegal.

Activities

(a) Clearly define each of the following terms:
- Sustainable economic activity
- The Kyoto Protocol
- Agriforestry
- Fishery exclusion zones.

(b) *'The sustainable use of resources provides a model for future economic development.'* Comment on the above statement with reference to Ireland's attitude to the use of fossil fuels.

(c) With reference to one example within and one example outside Europe, describe attempts to use forests in a sustainable way.

(d) Evaluate the extent to which developed countries such as Ireland contribute to the managed exploitation of sea fish stocks.

Chapter 13
Fair Trade

Find out more about fair trade in *Our Dynamic World 2* Chapter 7.

The globalisation of world trade has created much employment in China, Brazil and other parts of the Majority World, but as discussed in Chapter 10, globalised trade has also tended to increase inequalities between rich and poor and has led to exploitation and misery within many Third World communities.

The problem is that at present trade tends to be driven almost solely by the desire to make ever greater profits for a select few, rather than by the need to achieve human, community and environmental development for all. Fair globalised trade could play a major role in achieving the sustainable development of peoples of all nations.

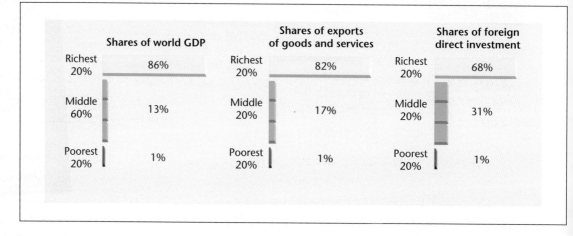

Fig. 13.1 Disparities in trade and investment. Describe the disparities in global opportunity shown here and outline some likely effects of these disparities on human development.

The FAIRTRADE Mark.

Working for Fairer Trade

The Fairtrade Mark Ireland and other Fairtrade labelling organisations have been set up to promote fairer international trade. Products that meet standards set by Fairtrade Labelling Organizations International (FLO) can be awarded the FAIRTRADE Mark. The **FAIRTRADE Mark** is a guarantee to consumers that the producers have been paid a fair price for the product in question, that the producers work in safe and decent conditions and that the production respects the local environment. There are now eighteen countries promoting Fairtrade and most of them use the new international FAIRTRADE Mark. Similar fair trade labels include the **Max Havelaar** label, which is used in the Netherlands, Belgium, France and Switzerland and the **Transfer** label, which is used in Germany, Austria and Luxembourg.

WHAT IS FAIR TRADE?

Fair trade might be taken to include the following elements:	The present situation:
1. The means to enable all producers and their families to earn adequate livings. This implies commodity **prices which are fair and which do not fluctuate too much** on the world market. It also implies long-term, fair relationships between the costs of First World and Majority World goods. These relationships must include **fair terms of international trade**. This ensures that prices paid for Third World products will be in some way index-linked to the prices charged by developed countries for the manufactured products that they sell to the Majority World.	The prices of Third World commodities such as coffee still fluctuate considerably on the world market. Multinational companies and First World countries have generally resisted the concept of index-linking the prices of Third World products to those of First World goods. A minority of buyers, however, follow a 'fair trade' policy and offer guaranteed minimum prices to producers.
2. **Fair wages and working environments** which do not harm producers either physically, psychologically or socially.	Some products sold in the West are made in countries such as China and Indonesia by people who work for very low wages in poor working environments.
3. Production which is both economically and ecologically **sustainable** – one which meets the needs of this generation without damaging the needs of future generations.	Some beef and hardwood (mahogany and teak) consumed in the developed world have been produced through the destruction of rainforests in countries such as Brazil and Indonesia.
4. **Fair access to international markets** for all countries and products. This would entail efficient methods of getting products from producers to consumers without involving the unnecessary use of speculators or trade 'middlemen'.	Powerful countries sometimes erect politically motivated trade barriers against weaker countries, even in contravention of international trade regulations. The United States has conducted such a trade blockade against neighbouring Cuba since 1961.
5. A reasonable degree of **control** for Third World governments over the resources of their own countries. This would imply some degree of control over First World multinational companies that operate in the South.	Trade globalisation and structural adjustment programmes by the International Monetary Fund have further reduced the control of governments over the activities of TNCs.
6. A preference for working with small scale producers and marginalised workers.	The dominance of powerful multinational companies over world trade is now greater than ever before. For example, seventy per cent of global trade is controlled by TNCs, and seventy per cent of the world's trade in wheat is now controlled by just six such companies.

How Fairtrade Works in Practice

Fairtrade can work in a simple but effective way, as is shown by the following example of Fairtrade involving some Nicaraguan coffee producers:

- The First World **importers** deal directly with the producers. This saves money because it eliminates the use of costly and unnecessary middlemen. The producers are offered a guaranteed minimum price which is higher than the international market price set on the New York coffee exchange. Some money is paid in advance, credit terms are arranged if needed and an extra premium is offered which can be used for local community projects.
- The **producers** agree to meet decent social standards for workers and environmental standards for the producing region.
- The **Fairtrade Labelling Organizations International** (FLO) certifies that the coffee has been produced meeting their standards.
- The **consumer** pays about two US cents extra per cup in the knowledge that the coffee they drink has been produced in a way that does not exploit the people or environment in the producing region.

Many consumers have shown themselves willing to pay a little extra for products that they know to have been produced and traded fairly. As a result, Fairtrade labels can now be found on some brands of coffee, cocoa, chocolate, honey, sugar, tea, fresh fruit and sportsballs. Their combined international sales account for more than €200 million. FAIRTRADE Mark products can be found in specialist Fairtrade shops, such as Oxfam Ireland shops, and increasingly through most of the main supermarkets.

How Fair Trade Creates Self-Reliance and Sustainable Development

Case Study 1: A Multipurpose Co-operative in the Philippines

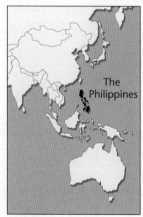

Fig 13.2

Barcelona is the name of a small area in the southernmost province of Luzon, the main island of the **Philippines**. The people of Barcelona were traditionally skilled in the production of hats, baskets, gift boxes and other craft items, which were traditionally sold to middlemen at very low prices.

In 1985, a British fair-trade organisation called **Traidcraft** first became involved in Barcelona's craft production. It provided training in dyeing, colour selection, design and pricing. Traidcraft also offered fair prices for the crafts produced and paid half of the total agreed price in advance.

This prepayment enabled people in Barcelona to set up a small producing group with twelve members. As the orders and income grew, the number of producers soon grew to fifty and the group was registered as a multipurpose co-operative with the aim of satisfying many basic needs of the local community. A small store and canteen were set up. Carpentry, metal craft and woodwork activities were organised. Forty per cent of the co-op's profits were set aside for a childcare centre. In Barcelona, health care was poor and there was only one doctor for 14,000 people. The co-operative organised training courses for local people in the techniques of basic herbal healing. It also organised the planting of suitable herbs so that an adequate supply would be readily available.

By facilitating the establishment of the Barcelona co-operative, fair trade helped poor people to provide for their own basic needs and to take control of their own lives. It thus assisted in the sustainable economic and social development of the Barcelona area.

Case Study 2:
An Irish Company Involved with Fairtrade

Bewley's of Dublin (now part of the Campbell Bewley Group) has been involved in the coffee-importing and café business since 1840. The Bewley name has long been associated with quality products and service, originally throughout Ireland and now also in Britain and the United States. While some commercial enterprises build their 'successes' on exploitation and sharp business practices, the Bewley family chose to build theirs on **ethical principles**. Their business aims include the following principles:

- To deliver excellent quality *products and services* at good value to customers
- To pay their *workers* fairly and to give each employee an opportunity to reach his/her full potential
- To pay coffee *producers* a fair price for their produce
- To always act responsibly on *environmental* matters
- To provide an attractive return for investors, while ensuring *sustainable future growth*.

Two sides to fair trading. Members of this co-operative in Nicaragua (left) provide coffee directly for Bewley's café in Dublin (right). Describe the nature of Bewley's relationship with its coffee producers.

Fig 13.3

In 1996, Bewley's became the first Irish company to be awarded the **FAIRTRADE Mark** for its '**Bewley's Direct**' coffee, so called because it is bought directly from coffee co-operatives in Costa Rica at fair prices. Producers for Bewley's receive a guaranteed minimum price and receive 'premiums' over and above the market purchasing price. These premiums are used, with local consultation, to fund basic-needs projects in the coffee-producing areas. Such a system, said Patrick Bewley, *'allows all those who consume and serve coffee to directly improve the lives of coffee producers who often find themselves in a constant battle against poverty.'*

The growing sales for Bewley's Direct Coffee over the years has shown that ethical business principles and Fairtrade can lead to sustainable and profitable economic growth. Over five per cent of Bewley's coffee is now sold on Fairtrade terms.

Activities

1. Do you consider that most of today's trading activities fall into the category of fair trade? Explain your point of view.
2. With reference to one area or trading example that you have studied, explain how fair trade can be made to operate in practice.
3. To what extent can fair trade be used as a means of stimulating sustainable human and economic development? Refer in your answer to actual examples of Fairtrade that you have studied.

CHAPTER 14
JUSTICE AND MINORITY GROUPS

The level of development of any society can be measured by the degree of justice with which that society treats its weakest members. These members are usually poor and are often members of identifiable and distinct minority groups or cultures. Such minority groups include *Palestinians* within Israel, *Amerindian peoples* such as the Yanomami in Brazil and the *Roma* ('gypsy') people of Romania. This chapter will focus in particular on justice issues relating to *Travellers* in Ireland and to *bonded labourers* in southern Asia.

Case Study 1: Travellers in Ireland

Who Are the Travellers?

There are about 22,550 members of the Traveller community. Travellers are native Irish people with a separate identity, culture and history from the majority of people in Ireland:

- **Nomadism** is a central characteristic of the Traveller community. This means that many Travellers earn their living and carry out important social and cultural activities in a way that requires frequent, planned movement. Members of an entire extended family may leave their usual halting places at any time and come together to be closer to a sick relative or to celebrate a wedding.
- Strong **family life and religion** are important to Travellers. Extended family loyalty is generally much more important among Travellers than it is among so-called 'settled' people.
- **Music** plays an important part in Travellers' lives and many Travellers are highly skilled musicians and singers. A distinctive Travellers' **language** called 'cant' still survives. **Traditions of arts and crafts** are also strong and these skills are handed down for generations within families.

Many Travellers who live in trailers or caravans remain in one place for the winter so that the children can attend school. They use the fine summer period to move from place to place to work as casual traders.

In Search of Justice

To our shame, Ireland's Travellers were recognised in 1991 by a European Parliament Committee of Inquiry as *'the single most discriminated against ethnic group'* within the European Union. Ireland's Economic and Social Research Institute (ESRI) had earlier concluded that *'the circumstances of the Irish Travelling people are intolerable. No humane or decent society, once made aware of such circumstances, could permit these circumstances to persist.'*

Injustices	Moves Towards Justice

Racism

Travellers have long suffered from **prejudice and racism**. An ESRI report from 1986 revealed that *'Irish Travellers are ... often despised and ostracised.'* A survey in 2000 found that forty-two per cent of Irish people admitted to holding negative views towards the Traveller community. It also found that fifty-nine per cent would not welcome a Traveller as a next-door neighbour, and that ninety-three per cent would not accept a Traveller as part of his or her family.

It is essential that the law and its enforcement should be used to stamp out racism against our Traveller minority. Limited progress has already been made in this regard:

- **The Prohibition of Incitement to Hatred Act** of 1991 forbids incitement to hatred of Travellers. By the year 2000, however, not a single case had been successfully brought under this Act.
- **The Equal Status Legislation** of 2000 offers most hope in the struggle against anti-Traveller discrimination. This law outlaws discrimination in the provision of all services. It means that a Traveller refused service in a bar or hotel, for example, can seek retribution in the courts. Travellers' groups have complained that a shortcoming of the legislation is that it requires complaints to be filed in writing. This discriminates against those Travellers who possess a limited formal education.

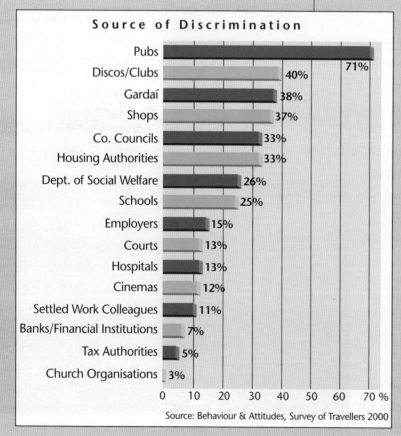

Fig. 14.1 Sources of discrimination against Travellers in Ireland.
Comment on the information provided here. Are you surprised by the information? Explain.

| Injustices | Moves Towards Justice |

Education

- There has been a tradition in some primary schools to **segregate** Traveller children into special classes, sometimes with separate lunchtimes from other children. Traveller support groups such as *Pavee Point* believe that this practice stemmed from the hostility of some settled parents to the inclusion of Travellers in their children's classes. The practice not only discarded the opportunity of having Traveller and settled children learn from each other's traditions, but also contributed to mutual distrust between the two communities.
- Education was sometimes used in the past to **'take the Traveller out of the child'**. This created cultural confusion, inferiority complexes and resentment among many Traveller children.
- While almost all Traveller children attend primary schools, only about 1,000 such children attend second-level schools at any one time. Only one-tenth of these proceed to enter fifth year, reflecting a tradition among many Travellers to see fifteen as the age at which formal education ends. Fewer than twenty Travellers attend third-level colleges each year. This **relative lack of second and third-level education** contributes to a widening prosperity gap between Travellers and settled people.

- In 1995, a government White Paper called for the full participation in school life by Traveller children by means of fully integrated classes which respected Traveller culture. This White Paper marked the beginning of a new, much-needed **intercultural educational policy**. Such intercultural education, which research shows is favoured by eighty-two per cent of the Irish public, can greatly help to reduce suspicions and misunderstandings between the Traveller and settled communities.
- There has also been a modification of the education system to meet the **special needs** of Travellers. The Department of Education and Science funds a visiting service in which teachers act as intermediaries between the Traveller communities and the school. These teachers visit halting sites and try to encourage Travellers to stay in school.

> 'I first became aware that being a Traveller meant being treated differently when I went to school. I was put in a Traveller-only class. We were also given a separate playtime. Our teacher looked upon us as a problem. She told us we were settled people 'gone wrong'. She told us that if we moved into houses we'd be 'all right again.'
> Mr P McDonagh, February 2001.

Accommodation and Living Conditions

By 1998, twenty-four per cent of Travellers still lived in **unserviced sites** by the sides of roads:
- These sites have no running water, toilet facilities, refuse collections or electricity. Such conditions *reduce life expectancy* among Travellers, only two per cent of whom live to sixty-five years or more.
- They result in *untidy and unsanitary conditions*, which in turn creates hostility towards Travellers on the part of the settled community.
- The poor living conditions that they provide make it almost impossible for Traveller children to succeed in *school*.

It is essential that Travellers' accommodation needs be met to a much fuller degree than currently pertains. The **Housing Act** of 1988 gives state recognition of the need for Traveller-specific accommodation. Key features of Traveller culture must be respected in whatever provisions are made. These features need to take account of the fact that extended families often live together and that Traveller families tend to be larger than the national average. It must also be remembered that some degree of nomadism is an important part of Traveller tradition. For these needs to be met a range of options including well-serviced halting sites, group housing schemes and conventional housing must be made available to our Traveller community.

| Injustices | Moves towards Justice |

Health

Infant mortality among Travellers is three times higher than that among the settled community, while **life expectancy** among adult Travellers is at least ten years less than among the average adult of the majority community.

- For the health of Travellers to improve significantly, their **living conditions** will have to be upgraded considerably.
- The creation of **national medical cards** for Travellers would also greatly contribute to improved Traveller health. A medical card is at present valid only for a single health board area. This means that if a Traveller moves – or is moved – to another area, a new medical card will have to be applied for and a new general practitioner found to accept the Traveller as his/her patient.
- The creation of **medical liaison officers** with Travellers would also be of benefit. These officers could visit Travellers' halting sites, informing people on sickness prevention and advising people on the health services available to them.

Travellers from the Pavee Point support group discussing the importance of primary health care with other Travellers.
Travellers themselves now play leading roles in organisations that work to achieve full empowerment and equality for the Traveller community in Ireland.

Case Study 2: Bonded Labour in Pakistan

According to the United Nations there are approximately twenty-seven million slaves in the world today, which is probably more than at any other time in history. Slaves can be described as people whose movement and choice of work is severely restricted and who do not have any bargaining power with the people who 'employ' them.

Bonded labour exists in several Third World countries and is the most common form of modern slavery. It affects up to twenty million people worldwide and approximately four million in Pakistan alone.

How Bonded Labour Works

(1) A poor person in the Third World may urgently need some money, perhaps to purchase medicine for a sick child. Such a person will find it almost impossible to obtain loans from banks, so he/she may resort to borrowing from an unscrupulous moneylending landlord or business person. The borrower 'bonds' (agrees) to work for the moneylender until the loan is paid off.

(2) Bonded labourers often lack even basic formal education and have no idea how interest is calculated. Wages paid by the moneylender may be so low and the rates of interest may be so high that the loan can never be paid off through the borrower's labour. The bonded labourer is now persuaded that he/she must work indefinitely and without wages for the borrower. Other members of his/her family may also have to work against the loan, which may even be handed down from generation to generation. In 1984, a poor family in Pakistan took a loan which was the equivalent of €90. By 2001, this 'debt' had grown to the equivalent of €890, despite the fact that a family of six had spent fifteen years working to pay it off.

The Search for Justice

In 1992, the Pakistani Parliament banned the practice of bonded labour within its jurisdiction. To date, however, the enforcement of this Act has not been vigorously pursued by the **government**:

- Moneylending landlords or business people are usually influential within their localities, and few of them are ever charged with using bonded labour.
- The Pakistani government has been in official denial of the level of bonded labour within its state. The government claims that no more than 7,000 bonded labourers exist within Pakistan. Human rights organisations estimate that the figure is close to four million.
- Pakistan has no system of identifying and providing alternative work for those bonded labourers who have been fortunate enough to be released from their slavery.

Voluntary bodies, such as the *Bonded Labour Liberation Front* of Pakistan, have battled to liberate bonded labourers and to hold governments and slave holders accountable for the abuses of slavery. The interventions of such groups have led to High Court rulings which have resulted in 12,000 bonded labourers being freed over the past ten years.

The Catholic Bishops' Conference of Pakistan set up the *National Commission for Justice and Peace (NCJP)* in 1984. This human rights organisation has been to the fore in the struggle against bonded labour in the brick industry. The NCJP uses seminars and newspaper articles to create public awareness of the issue. It also launched a campaign of direct solidarity with victims and operates a networking process with other nongovernmental organisations. These campaigns bore fruit when the Lahore High Court outlawed the advancement of money in return for bonded labour in the brick industry. As a result, the debts of thousands of bonded labourers were written off.

Iqbal Masih – Child Labourer and Campaigner

When Iqbal's parents took a bonded loan of less than €18, Iqbal was enslaved at a carpet loom in Pakistan from the age of four until the age of ten. He was tied to his loom, tying tiny knots for twelve hours a day, every day. Six years later, when he went to his boss to demand freedom, he was told that his debt had grown to approximately €470.

Iqbal was freed with the help of the *Bonded Labour Liberation Front* and soon became president of the youth wing of that movement. He travelled to Europe and the United States, urging support against the slavery of bonded and child labour in the carpet industry.

Back in Pakistan, he received several death threats in the spring of 1995. On 16 April of that year, he was shot dead while out cycling with two relatives. The Pakistani government reported his death as accidental.

Iqbal's murder caused a rapid growth in the international campaign to end bonded labour in the carpet making industry. The RUGMARK label is your best assurance that no illegal child labour has been used to make your carpet or rug. The success of the brand in Europe has resulted in the freeing of thousands of slaves.

Bodies such as the NCJP call for people from outside Pakistan to help in the struggle against bonded labour. They point out that while some slave owners enjoy considerable influence within the country, Pakistan's government is sensitive to international pressures on it to enforce its own laws.

In response to these appeals, international bodies such as the *United Nations High Commission for Human Rights* (Geneva), *Antislavery International* (London) and *Save the Children* (London) have all joined the fight against bonded labour worldwide. In 2001, the Irish NGO *Trócaire* embarked on the following three-fold Lenten Campaign to oppose modern slavery:

1. It supported local groups such as the NCJP in the fight against bonded labour.
2. It paid for education, training and grants so that freed bonded labourers might be offered a fresh start in life.
3. It urged Irish people to join the campaign for an end to all forms of modern slavery (see picture on the next page).

JUSTICE AND MINORITY GROUPS

In the brick-making industry, children who are bonded labourers must perform back-breaking work and young girls are sometimes sexually abused.
This picture is of a postcard issued free to Irish people by Trócaire as part of its 2001 Lenten Campaign. The other side of the card is addressed to the Irish Minister for Enterprise, Trade and Employment and requests the Irish government to use its influence to highlight the injustice of bonded labour.

Activities

1. Write a short paragraph on each of the following terms:
   - Nomadism
   - The Equal Status Legislation Act (Ireland)
   - Bonded labour
   - The Bonded Labour Liberation Front
   - The National Commission for Justice and Peace (Pakistan).
2. (a) Identify four minority groups outside of Ireland that you feel are not enjoying full human rights within the country or countries in which they live.
   (b) In the case of any one of the groups that you identify in (a), describe the injustices suffered by the group and evaluate efforts being made to combat these injustices.
3. *'Sustainable development in Ireland must incorporate improved levels of justice for minority groups within our country.'* Discuss this statement with reference to any minority group in Ireland.

CHAPTER 15
SELF-HELP DEVELOPMENT

Economic globalisation has concentrated ever-increasing wealth and power in the hands of fewer and fewer people. A small number of individuals and corporations have become so powerful that the self-reliance of local communities has become seriously limited and even the powers of national governments have become eroded. Such changes have increased poverty and powerlessness among the many poor people of our planet.

Some geographers believe that only the decentralisation of economic and political influence can empower poor people and force more governments and lawmakers to put human concerns before economic growth. Such people-centred development is often best achieved by groups that promote self-reliance and **self-help at local level**.

Self-help groups are typically equally open to all and are answerable to their members. Being 'of the people and for the people' of their localities, they represent democracy in its most direct form. Such groups often form spontaneously in answer to pressing local crises or needs and some survive and develop into long-standing movements that offer alternatives to existing dominant power structures. Some local groups link up with and receive assistance from **larger development bodies**, such as the following:

- *Nongovernmental organisations* (NGOs), such as Trócaire, Oxfam and Christian Aid. These bodies often possess expert knowledge on matters affecting human rights, asylum and social and health issues. They may also be in a position to financially help the development of self-help groups, especially in their early stages of growth.
- *The women's movements*, which in various places work for goals that would benefit all marginalised people. These goals may include the abolition of discrimination in education, wages and career opportunities along with the promotion of public childcare and guaranteed basic salaries.
- *Cultural movements* which, especially in the South, work to strengthen local cultures as alternatives to the relentless encroachments of Westernised norms and values.

Relate the message of this cartoon to the subject of self-help.

SELF-HELP IN ACTION

Case Study 1: Self-Help Houses in Lusaka, Zambia

The Zambian government cannot afford large public housing schemes for the many homeless people who have flocked from rural areas into Lusaka, the country's capital. Instead, the government and the World Bank together have encouraged local communities to organise the building of their own homes.

When a small group of twenty to thirty local people band together to organise homebuilding, they are given a plot of land of approximately ten hectares with a water tap on it. The World Bank and the government then provide enough money to build simple homes and to lay water and drainage pipes, provided that the group itself digs the trenches and foundations needed. If the locals also build the shells of the houses, the money saved is used to add electricity and tarmac for the roads. A school or clinic might eventually be added using the same self-help formula.

Schemes such as this are a good example of how self-reliant groups can combine with outside bodies to achieve local development in an inexpensive manner that is *appropriate* to their resources and needs. The local people not only acquire homes for themselves, they also develop self-esteem, community spirit and useful skills.

Fig.15.1

Trócaire has assisted this self-help building project, in which several of the builders are women.

Case Study 2: Self-Reliance in Kerala, India

The people of Kerala in southern India (see Figure 15.2) live on average more than ten years longer than Indians in general. They enjoy better health, education and transportation services, as well as lower inequalities between the sexes or castes than people in other Indian regions. The activities of 'grassroots' self-help groups, supported by a wise regional government, have been largely responsible for this level of development and, in particular, for the following developmental activities:

- **Land reform** gave more than one million people direct access to land. Co-operative credit was arranged to allow people to purchase essential livestock and inexpensive hand tools. This allowed rural people to increase their incomes as well as their sense of self-reliance and self-worth.
- Community-based schemes helped to provide **clean water and primary health care** to many villages and rural areas. These projects combined to greatly increase public health in the region and to reduce disease and death rates among the people. This, in turn, helped to improve economic output in the region.
- As people became more healthy and economically secure, their need to have large families was reduced. Locally run **family-planning campaigns** helped them to reduce birth rates and population pressures in the region.

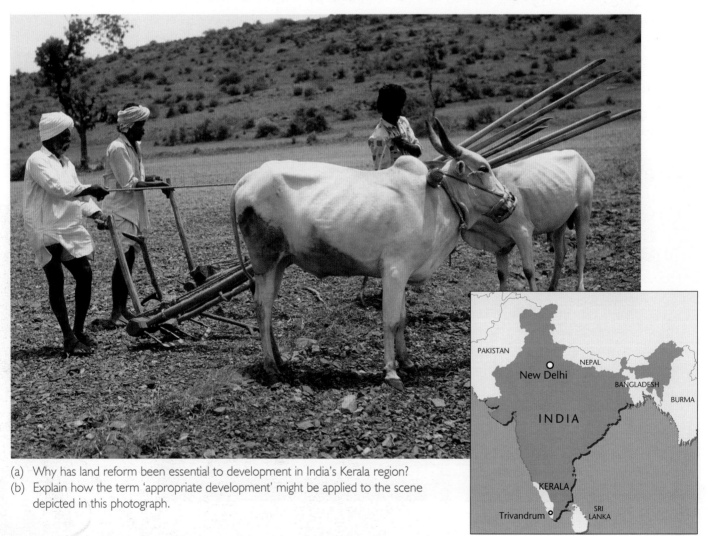

(a) Why has land reform been essential to development in India's Kerala region?
(b) Explain how the term 'appropriate development' might be applied to the scene depicted in this photograph.

Fig. 15.2

Case Study 3: Self-Help in the Irish Home

Global warming is widely regarded as one of the most worrying environmental problems of our time, while *waste disposal* poses a serious national challenge for Ireland. Yet even these large-scale problems could be considerably reduced by greater degrees of self-reliance within individual households in Ireland and other Western countries. Each family could become a self-help environmental unit by adopting tactics such as those outlined in the text boxes below.

Global Warming

To help combat global warming, simply use less energy. Adopt as many of the following policies as possible:

- *Conserve domestic energy.* Use energy-efficient lights and appliances. Use thermostats and timers on heating systems. Lag hot water tanks and always turn off unused lights and appliances. Insulate your loft and walls, and make doors and windows draught-proof.
- When burned, natural gas emits less carbon dioxide than coal or oil. Carbon dioxide is a greenhouse gas, so where feasible, *use natural gas* in preference to coal or oil.
- Use public *transport* in preference to private vehicles. This helps to reduce fuel consumption as well as traffic congestion. Where feasible, walk or cycle to work. This saves energy, money and stimulates physical fitness. If you do take your car to work, try to organise a car pool with other workers.

Domestic Waste

The average Irish home now produces almost one and a half tons of domestic waste each year. *Recycling* is a more effective solution than disposal to our waste problem. More than half of the waste we produce could be usefully recycled instead of being incinerated, buried or dumped. Recycling is also good for the economy because it reduces energy costs and lessens the amounts of raw materials being used.

Some waste food can be recycled into useful compost for gardens. Glass, paper and metals can be left at separate public authority recycling bins which now exist in car parks or near supermarkets in many urban areas. In 2001, Dublin Corporation provided each household with three different wheelie bins in which domestic waste could be separated to facilitate its recycling.

Public authority recycling bins in Dublin. Why is the use of these bins an important form of self-help?

Activities
1. Explain the meaning of the term 'self-reliant groups' and, with reference to some examples, evaluate the roles of such groups in Third World development.
2. Examine the news article in the text box below.
   (a) Outline the negative effects of globalisation referred to in this article.
   (b) Suggest some practical ways in which the concept of self-reliance might improve the situations of the people who suffer from these negative effects.

Economic Globalisation and Zambia

People are dying, quietly but in huge numbers, all over Zambia. They die, not because of some accident of nature, but as a direct result of economic policies imposed by faceless Western planners. For over twenty years the World Bank and the International Monetary Fund have been forcing structural adjustment programmes (SAPs) on the bankrupt countries of Africa, blind to the havoc they are causing.

Enter the children's ward of Siavinga Hospital in Southern Zambia and the smell hits you like a wall – a musty, medicinal odour. Rows of poor children lie on small beds, slowly passing away from preventable diseases like TB, malaria, and pneumonia.

On the other side of the building is a cleaner, neater ward where half of the beds stand empty. This is the fee-paying section, where families can buy a better chance of life. In World Bank language, this is 'user responsive healthcare'.

Meanwhile, in the nearby shanty town of Misisi, Masauso Phiri nearly died of starvation – luckily he was saved by the return of his son from the copper mines. Mr Phiri has heard of structural adjustment. It was because of structural adjustment that he lost his job as a security guard and one of his children – a three-year-old boy who died of pneumonia.

Structural adjustment has, of course, stimulated some foreign investment in Zambia. But all investment is not necessarily good. Take Shoprite, the South African supermarket chain that is colonising the country with the help of a massive government tax rebate. Shoprite has laid waste the economies of entire towns, undercutting local traders and putting stores run for generations by one family out of business. To make matters worse, Shoprite buys nothing locally. Their tax-free produce – even maize and potatoes – is trucked in from Zimbabwe and South Africa. Meanwhile Zambian maize rots in the fields, because the farmers who grow it cannot find a market.

(Mark Lynas, *New Internationalist*, January/February 2000)

Picture Credits

The author and publisher gratefully acknowledge the following for permission to reproduce photographic material:

PHOTOS
ANTHONY BLAKE PICTURE LIBRARY: 26 © Maximilian Stock Ltd/Anthony Blake Picture Library
BORD NA MÓNA: 2 © Sceal na Mona
CAMERA PRESS: 44, 77, 91L © Camera Press
CORBIS: 3L © Jeremy Horner; 5 © J.B. Russell; 12B, 16 © David Turnley; 14L © Peter Turnley; 19B © Jonathan Blair; 31R © Collart Herve/Corbis Sygma; 34 © Sally A. Morgan; 58L © Chris Rainier; 58R © Liba Taylor; 61 © Carrion CA/Corbis SYGMA; 67 © Sean Sexton; 75 © Todd Haimann; 91R © Bernard and Catherine Desjeux; 95R © Inge Yspeert; 102 © John Van Hasselt
DEREK SPEIRS: 49, 97, 100 © Derek Speirs
EXILE IMAGES/IRISH REFUGEE COUNCIL: 50 © Howard Davies
FAIRTRADE: 92 © Fairtrade
FINBARR O'CONNELL: 25 © Finbarr O'Connell Photography
THE HUTCHINSON PICTURE LIBRARY: 38 © Sarah Errington; 81T © R. Ian Lloyd
LOUTH LEADER: 71B © Louth Leader
PANOS: 7 © Chris Sattlberger; 39 © J.B. Russell; 53, 65 © Crispin Hughes; 103 © Panos/ Trócaire
PA PHOTOS: 82 © Westh Haydn West/ PA Photos
PHOTOCALL IRELAND: 107 © Gareth Chaney
PHOTO IMAGES: 71T © Photo Images Ltd
REX FEATURES: 31L © Sipa Press
REUTERS: 90 © Mike Brown/ Reuters
ROBERT HARDING: 106 © David Lomax
SCIENCE PHOTO LIBRARY: 4 © Maximilian Stock Ltd/ Science Photo Library
SOUTH AMERICAN PICTURES: 3R, 19L © Tony Morrison
STILL PICTURES: 12T, 95L © Jorgen Schytte; 14R © Hartmut Schwarzbach; 19R © Dominic Harcourt-Webster; 42 © Mark Edwards; 63 © Ron Gilling; 74 © Harmut Schwarzbach;
TRÓCAIRE: 105 © Trócaire
UNICEF: 27 © Dr M. Khan/ UNICEF

CARTOONS
BRICK (www.brickbats.co.uk): 16T, 16B, 28, 56T, 56B, 76, 78, 83 © Brick
CARTOONISTS & WRITERS SYNDICATE/ CARTOONWEB.COM: 7 © McDonald; 81B © Plantu
THE ECOLOGIST: 8 © Richard Willson
TIDE, Development and Education Centre: 104 © TIDE, Development and Education Centre
MARTYN TURNER: 10, 61 © Martyn Turner
THE WASHINGTON POST WRITERS GROUP: 13 © Signe Wilkinson/ 1992, The Washington Post Writers Group

Map extract reproduced by permission of Ordnance Survey, Ireland.

The author and publishers have made every effort to trace all copyright holders, but if any has been inadvertently overlooked we would be pleased to make the necessary arrangements at the first opportunity.